Introduction to Nanoscience and Nanotechnology

Introduction to Nanoscience and Nanotechnology

Adriel Dawson

STATES
ACADEMIC PRESS
www.statesacademicpress.com

Published by States Academic Press,
109 South 5th Street,
Brooklyn, NY 11249, USA

Copyright © 2022 States Academic Press

This book contains information obtained from authentic and highly regarded sources. All chapters are published with permission under the Creative Commons Attribution Share Alike License or equivalent. A wide variety of references are listed. Permissions and sources are indicated; for detailed attributions, please refer to the permissions page. Reasonable efforts have been made to publish reliable data and information, but the authors, editors and publisher cannot assume any responsibility for the validity of all materials or the consequences of their use.

Trademark Notice: Registered trademark of products or corporate names are used only for explanation and identification without intent to infringe.

ISBN: 978-1-63989-309-6

Cataloging-in-Publication Data

   Introduction to nanoscience and nanotechnology / Adriel Dawson.
     p. cm.
   Includes bibliographical references and index.
   ISBN 978-1-63989-309-6
   1. Nanoscience. 2. Nanotechnology. 3. Nanostructures. I. Dawson, Adriel.
T174.7 .I58 2022
620.5--dc23

For information on all States Academic Press publications visit our website at www.statesacademicpress.com

STATES
ACADEMIC PRESS

Contents

Preface VII

Chapter 1 **Understanding Nanoscience and Nanotechnology** 1
    i.   Nanoscience 1
    ii. Nanotechnology 4
    iii. Nanomaterials 4
    iv. Nanoscale 10

Chapter 2 **Nanoparticles: Synthesis and Application** 20
    i.   Characterization of Nanoparticle 20
    ii. Structure of Nanoparticles 26
    iii. Energy Bands 35
    iv. Physicochemical Properties 37
    v. Surface Area of Nanoparticles 43
    vi. Basic Optical Properties of Nanoparticles 48
    vii. Synthesis of Nanoparticles 60
    viii. Silver Nanoparticles 66

Chapter 3 **Quantum Dots** 82
    i.   Core Structure of Quantum Dots 82
    ii. Working of Quantum Dots 83
    iii. Properties of Quantum Dots 85
    iv. Types of Emission 88
    v. Preparation of Quantum Dots 92
    vi. Applications of Quantum Dots 103

Chapter 4 **Understanding Nanostructure** 110
    i.   Nanowire 111
    ii. Nanomesh 120
    iii. Nanohole 123
    iv. Nanosheet 124
    v. Nanopillar 125
    vi. Nanostructured Film 127

	vii.	Sculptured Thin Film	128
	viii.	Gradient Multilayer Nanofilm	129
	ix.	Icosahedral Twins	130
	x.	Nano Flake	132
	xi.	Nanofoam	133
	xii.	Thermodynamics of Nanostructures	134

Chapter 5 Diverse Aspects of Nanotechnology 142
    i. Vapor-Liquid-Solid Method 142
    ii. Nanopore Sequencing 149
    iii. Molecular Self-assembly 155
    iv. DNA Nanotechnology 157
    v. Self-assembled Monolayer 171
    vi. Supramolecular Assembly 179

Chapter 6 Carbon Nanotubes 183
    i. Types of Carbon Nanotubes 183
    ii. Synthesis of CNT 191
    iii. Functionalization of Carbon Nanotubes 194
    iv. Properties of CNT 196
    v. Applications of CNT 199

Chapter 7 Consequences of Nanotechnology 203
    i. Impact of Nanotechnology 203
    ii. Societal Impact of Nanotechnology 209
    iii. Nanotoxicology 214
    iv. Nanomaterials Pollution 223
    v. Regulation of Nanotechnology 225

Permissions

Index

Preface

The branch of science which focuses on the study and manipulation of structures and materials on the nanometer scale is called nanoscience. It is a multi-disciplinary field which employs the principles of physics, chemistry, medicine, biology, material science, engineering and computing. It seeks to develop an understanding of the mechanical, electrical and optical properties of these structures, as they are different from the macro-scale properties due to the quantum mechanical effects. Nanotechnology refers to the industrial application of matter on an atomic, molecular, and supramolecular scale. It draws on the principles of various other branches such as organic chemistry, surface science, semiconductor physics, molecular biology, micro-fabrication, and molecular engineering. Nanotechnology finds extensive application in the areas of medicine, electronics, consumer products and energy production. This book provides comprehensive insights into the field of nanoscience and nanotechnology. It presents the complex subject of nanoscience and nanotechnology in the most comprehensible and easy to understand language. Those with an interest in the field of nanoscience and nanotechnology would find this book helpful.

A short introduction to every chapter is written below to provide an overview of the content of the book:

Chapter 1 - The study of structures and materials on nanoscale is known as nanoscience. The science and technology which makes use of matter on a nanoscale to design and produce structures, devices and systems at the nanometer scale is known as nanotechnology. This is an introductory chapter which will introduce briefly all the significant aspects of nanoscience and technology; **Chapter 2** - The particles which have dimensions between 1 and 100 nanometers are called nanoparticles. The properties of nanoparticles are remarkably different from the larger particles of the same substance. Nanoparticles are often used to improve the properties of a material. This chapter has been carefully written to provide an easy understanding of the varied facets of nanoparticles; **Chapter 3** - The man-made nanoscale crystals which have optical and electronic properties which differ from larger particles are known as quantum dots. When quantum dots are exposed to UV light, electrons in the quantum dots get excited to a state of higher energy. This chapter closely examines the key concepts of quantum dots to provide an extensive understanding of the subject; **Chapter 4** - Any structure with at least one dimension measuring in nanometer range is known as nanostructure. This chapter discusses various nanostructures such as nanowire, nanomesh, nanohole, nanosheet, nanopillar, nanostructured film, sculptured thin film, gradient multilayer nanofilm, etc. All the diverse principles of nanostructures have been carefully analyzed in this chapter; **Chapter 5** - The mechanism which is used to grow one-dimensional structures like nanowires is vapor-liquid-solid method. This method makes use of foreign element catalytic agent to mediate the growth. This chapter covers fundamental concepts of nanotechnology such as nanopore sequencing, molecular self-assembly, DNA nanotechnology, self-assembled monolayer and supramolecular assembly; **Chapter 6** - The cylindrical molecules made of sheets of single-layer carbon atoms are known as carbon nanotubes. Carbon nanotubes have remarkable electrical conductivity. In order to completely understand carbon nanotube, it is necessary to understand the processes related to it. The following chapter elucidates the varied processes and mechanisms associated with this area of study; **Chapter 7 -** There are several impacts and benefits of nanotechnology such as water purification systems, nanomedicine, improved manufacturing methods, better food production methods, etc. This chapter covers societal impact of nanotechnology, nanotoxicology, nanomaterials pollution and regulation of nanotechnology. The topics elaborated in this chapter will help in gaining a better perspective about the impacts of nanotechnology.

I extend my sincere thanks to the publisher for considering me worthy of this task. Finally, I thank my family for being a source of support and help.

Adriel Dawson

Understanding Nanoscience and Nanotechnology

The study of structures and materials on nanoscale is known as nanoscience. The science and technology which makes use of matter on a nanoscale to design and produce structures, devices and systems at the nanometer scale is known as nanotechnology. This is an introductory chapter which will introduce briefly all the significant aspects of nanoscience and technology.

Nanoscience

Nanoscience is the study of processes and manipulation of materials at atomic or molecular scale, such that the properties vary considerably than at larger scales, i.e., bulk materials.

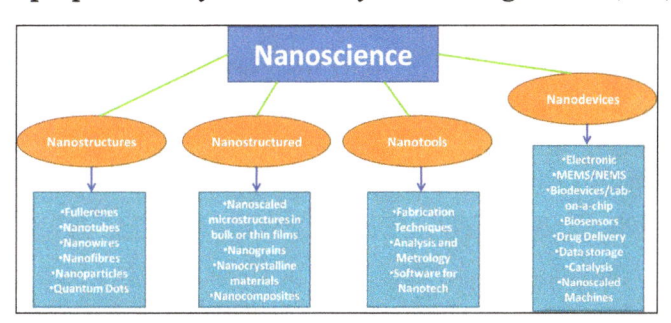

Bulk and micron-sized materials (such as a sand granule) demonstrate continuous (macroscopic) physical properties. However, as the size of the materials is decreased down to the nanoscale, classical physics fails to explain their behaviors or properties (e.g., energy, momentum, etc.), and quantum mechanics needs to be applied to describe them. For example, depending upon the particle size, gold demonstrates very different properties (optical, electrical, mechanical, etc.) at Nano scale.

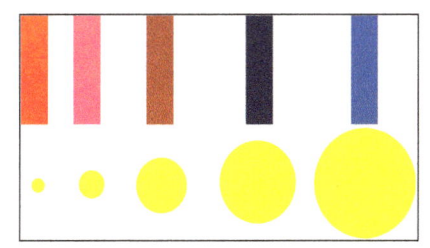

Figure: Different sized gold particles produce different colors at nanoscale.

Nanotechnologies include "nanoscale designing, characterizing, and producing structures, devices and systems by controlling shape and size for their applications in various fields".

Analogous to several other disciplines, nanotechnology was in use several centuries before any formal definition of the field. For example, nanotechnology was widely used in steel and painting industries. Early contributions came from James Clark Maxwell and Richard Adolf Zsigmondy. Zsigmondy investigated colloidal solutions including sols of gold as well as other nanomaterials. Irvin Langmuir and Katherine B. Blodgett contributed significantly to the field in first half of twentieth century.

American physicist Richard Feynman is credited with the first modern systematic discussion and formal announcement of nanotechnology as an important field of scientific endeavor. Though he did not coin the term 'nanotechnology', he emphasized the significance of manipulation of matter at very small scale such that these studies will allow the understanding of processes occurring in complex situations. He talked about size dependent behavior of various phenomena and proposed various challenges such as creating a nanomotor.

Norio Taniguchi, Japanese scientist, mentioned the term 'nanotechnology' for the first time in 1974 in his paper on synthesis technology to create objects and features of nanometer dimension. He writes:

"In the processing of materials, the smallest bit size of stock removal, accretion or flow of materials is probably of one atom or one molecule, namely 0.1–0.2 nm in length. Therefore, the expected limit size of fineness would be of the order of 1 nm Nano-Technology mainly consists of the processing separation, consolidation and deformation of materials by one atom or one molecule".

K. Eric Drexler, an American engineer, is often cited with the developer of molecular nanotechnology which led to the development of nanosystem machinery.

The respective discoveries of scanning tunneling and atomic force microscopes in 1980s are considered milestone for the development of nanotechnology as a filed. These microscopes allowed atomic level imaging of materials which is crucial for manipulating matter at atomic/molecular scales. The parallel advancements in computer technology facilitated large scale simulations and analyses of materials by supercomputers, thereby providing significant insights into the structure as well as properties of the materials. The simultaneous modeling, visualizing, as well as manipulation activities, greatly stimulated research investigations in the twentieth century.

The discovery of new carbon structure, Bucky ball made up of 60 carbon atoms, further increased the interest in the field. This propelled the investigation to discover other nanostructures made up of carbon as well as other materials. The 1991 finding of carbon nanotubes has been instrumental in enthusing research activities in the field of Nanoscience and nanotechnology; consequently, numerous nanotechnologies have been unearthed. Carbon nanotubes are one of the most studied materials owing to its amazing properties such as it gives ~100 times strength than steel at 1/6th of the weight. Similarly, novel structures including Quantum dots have been produced which demonstrate properties intermediary to the bulk and single molecular structures.

The astounding physical, chemical as well as mechanical behaviors of nanomaterials, such as large surface area, tiny size, etc.; offer extensive applications in numerous fields including electronics, optoelectronics, etc. Presently, molecular electronics is being proposed as new technological breakthrough. As silicon IC technology is approaching its physical limits, nanotechnology is expected to drive the future generation electronics. If devices can be made from a small cluster of atoms or molecules, computer chips can be fabricated to house 10 times more transistors than possible by the present technology.

Nanometre Scale

Conventionally, nanoscale has been defined as 1-100 nm. A nanometer is one billionth part of one meter. The minimum size is set to 1 nm, thus a single atom or very small clusters of atoms cannot be considered as nanoparticles. Consequently Nanoscience and nanotechnology involve a cluster of atoms at least 1 nm in size. However, the necessary condition for denoting clusters of atoms as nanomaterials is the onset of a quantum phenomenon instead of the actual dimension at which the effect occurs. Furthermore, above nanoscale, the properties of a material result from bulk or volume effects, namely the type of the atoms present, type of bonding existing between them, and the stoichiometry. Below this point, the properties of the materials change and even though the type of atoms and their organizational orientation are still important, surface area effects begin to dominate. The surface area effects include the size as well as the shape of the object.

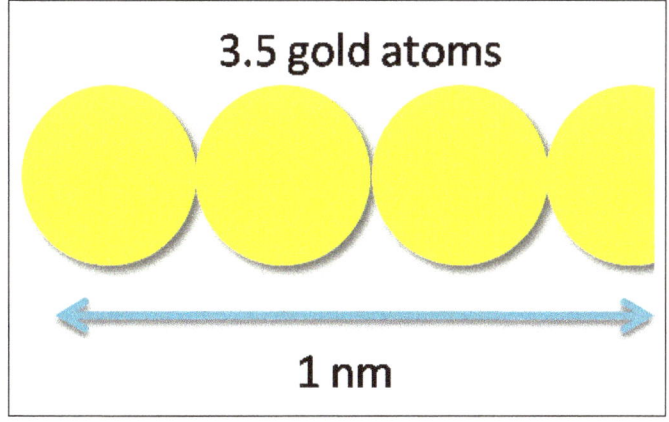

Figure: Three and a half gold atoms (each with the covalent radius of 0.144 nm) positioned adjacently in a row equal 1 nm.

Therefore, Nanoscience is not just the science of the small, but the science where materials with small dimensions demonstrate new physical phenomena, collectively described as quantum effects. Quantum effects are size-dependent and significantly differ from the properties of macroscaled or bulk materials. The changed properties may involve, though not limited to, color, solubility, strength, electronic and thermal conductivities, magnetic behavior, mobility, chemical and biological activities, etc.

Nanotechnology

Nanotechnology is the development and the use of techniques to study physical phenomena and develop new devices and material structures in the physical size range from 1 to 100 nanometres (nm), where 1 nanometre is equal to one billionth of a meter. Nanotechnology influences almost all areas of our lives, including materials and manufacturing, electronics, computers, telecommunication and information technologies, medicine and health, the environment and energy storage, chemical and biological technologies and agriculture.

Nanomaterials

A natural, incidental or manufactured material containing particles, in an unbound state or as an aggregate or as an agglomerate and where, for 50 % or more of the particles in the number size distribution, one or more external dimensions is in the size range 1 nm - 100 nm.

The objects with at least one dimension in nanometer range (i.e., ~1-100 nm) are termed as nanomaterials. Dimension dependent classification of nanomaterials is enumerated in Table below.

Dimension	Type	Example
All three dimensions < 100 nm	Nanoparticles, Quantum dot, Nano shell, nanoring, etc.	Bulk
Two dimensions < 100 nm	Nanotube, nano-fibres, nanowire, etc.	Bulk
One dimension < 100 nm	Thin film, layer or coating.	Bulk

Broadly, nanomaterials are classified into two types:

- "Non-intentionally produced nanomaterials", these are the Nano sized particles or materials which are present in the environment naturally (e.g. proteins, viruses, nanoparticles released from volcanic eruptions, etc.) or a result of human activities (e.g. particles emitted from fossil fuel combustions).

- "Intentionally produced nanomaterials" are the nanomaterials produced by following a specific synthesis technique. These are the desired nanomaterials which are produced for a particular application.

Nanotechnologies, as a matter of fact, do not include the non-intentionally produced nanomaterials.

Different Types of Nanoparticles

Nanoparticles can be classified into different types according to the size, morphology, physical and chemical properties. Some of them are carbon-based nanoparticles, ceramic nanoparticles, metal nanoparticles, semiconductor nanoparticles, polymeric nanoparticles and lipid-based nanoparticles.

Carbon-based Nanoparticles

Carbon-based nanoparticles include two main materials: carbon nanotubes (CNTs) and fullerenes. CNTs are nothing but graphene sheets rolled into a tube. These materials are mainly used for the structural reinforcement as they are 100 times stronger than steel.

CNTs can be classified into single-walled carbon nanotubes (SWCNTs) and multi-walled carbon nanotubes (MWCNTs). CNTs are unique in a way as they are thermally conductive along the length and non-conductive across the tube.

Fullerenes are the allotropes of carbon having a structure of hollow cage of sixty or more carbon atoms. The structure of C-60 is called Buckminsterfullerene, and looks like a hollow football. The carbon units in these structures have a pentagonal and hexagonal arrangement. These have commercial applications due to their electrical conductivity, structure, high strength, and electron affinity.

Ceramic Nanoparticles

Ceramic nanoparticles are inorganic solids made up of oxides, carbides, carbonates and phosphates. These nanoparticles have high heat resistance and chemical inertness. They have applications in photo catalysis, photo degradation of dyes, drug delivery, and imaging.

By controlling some of the characteristics of ceramic nanoparticles like size, surface area, porosity, surface to volume ratio, etc., they perform as a good drug delivery agent.

These nanoparticles have been used effectively as a drug delivery system for a number of diseases like bacterial infections, glaucoma, cancer, etc.

Metal Nanoparticles

Metal nanoparticles are prepared from metal precursors. These nanoparticles can be synthesized by chemical, electrochemical, or photochemical methods. In chemical methods, the metal nanoparticles are obtained by reducing the metal-ion precursors in solution by chemical reducing agents. These have the ability to adsorb small molecules and have high surface energy.

These nanoparticles have applications in research areas, detection and imaging of biomolecules and in environmental and bio analytical applications. For example gold nanoparticles are used to coat the sample before analyzing in SEM. This is usually done to enhance the electronic stream, which helps us to get high quality SEM images.

Semiconductor Nanoparticles

Semiconductor nanoparticles have properties like those of metals and non-metals. They are found in the periodic table in groups II-VI, III-V or IV-VI. These particles have wide band gaps, which on tuning shows different properties. They are used in photo catalysis, electronics devices, photo-optics and water splitting applications.

Some examples of semiconductor nanoparticles are GaN, GaP, InP, InAs from group III-V, ZnO, ZnS, CdS, CdSe, CdTe are II-VI semiconductors and silicon and germanium are from group IV.

Polymeric Nanoparticles

Polymeric nanoparticles are organic based nanoparticles. Depending upon the method of preparation, these have structures shaped like Nano capsular or Nano spheres. A Nano sphere particle has a matrix-like structure whereas the Nano capsular particle has core-shell morphology. In the former, the active compounds and the polymer are uniformly dispersed whereas in the latter the active compounds are confined and surrounded by a polymer shell.

Some of the merits of polymeric nanoparticles are controlled release, protection of drug molecules, ability to combine therapy and imaging, specific targeting and many more. They have applications in drug delivery and diagnostics. The drug deliveries with polymeric nanoparticles are highly biodegradable and biocompatible.

Lipid-based Nanoparticles

Lipid nanoparticles are generally spherical in shape with a diameter ranging from 10 to 100nm. It consists of a solid core made of lipid and a matrix containing soluble lipophilic molecules. The external core of these nanoparticles is stabilized by surfactants and emulsifiers. These nanoparticles have application in the biomedical field as a drug carrier and delivery and RNA release in cancer therapy.

Thus, the field of nanotechnology is far from being saturated and it is, as the statistic says, sitting on the staircase of an exponential growth pattern. It is basically at the same stage as the information technology was in the 1960s and biotechnology in the year of 1980s. Thus it can easily be predicted that this field would witness a same exponential growth as the other two technological field witnessed earlier.

Applications of Nanotechnology

Health Sector

- The application of nanotechnology in the health sector is wide-ranging.
- Nanomaterials can be used inside and outside the body.
- Thus, the integration of nanomaterials with biology has led to the development of diagnostic devices, analytical tools, drug delivery vehicles, and physical therapy applications.
- This technology has led to the possibility of delivering drugs to precise cells, ensuring greater efficiency and lesser side effects.
- Nanomaterials have also given the scope for repairing damaged tissues since the cells can be artificially produced using this technology.
- This technology has currently become an important diagnostic tool since it can sense and label specific molecules, structures or microorganisms.

Food Industry

- Nanotechnology provides the potential for safe and better quality food and improved texture and taste of the food.
- A contamination sensor, using a flash of light can reveal the presence of E-coli.
- Antimicrobial packaging made out of cinnamon or oregano oil or nanoparticles of zinc, calcium, etc., can kill bacteria.
- The Nano-enhanced barrier can keep oxygen-sensitive food fresh.
- Nano-encapsulating can improve the solubility of vitamins, antioxidants, healthy omega, etc.
- Nano-fibres made of lobster shells or organic corn can allow for antimicrobial packaging while being biodegradable.

- Nano barcodes are used to tag individual products and trace outbreaks.

Electronic Components

- Computers are already working on a nanoscale.
- Nanotechnology has greatly improved the capacity of electronic components by:
  - Reducing the size of the integrated circuits' transistors.
  - Improving the display screens of the electronic devices.
  - Reducing power consumption, weight, and thickness of the electronic devices.

Energy-efficient

- This technology can improve the efficiency of the existing solar panels. It can also make the manufacturing process of solar panels cheaper and efficient.
- It can improve the efficiency of fuel production and consumption of petroleum materials.
- It is already being made use of in many batteries that are less-flammable, efficient, quicker-charging and are lightweight and higher power density.
- Broadly, it has the potential to improve the existing technologies to be more efficient with less consumption of energy.

Textile Industry

- Nanotechnology has already made revolutionary changes in the textile industry and is estimated to make a market impact worth hundreds of billions of dollars.
- Nanoscience has now produced stain and wrinkle resistant cloths and may further improve upon the existing innovations.

Environment

- The nanotechnology has numerous eco-friendly applications.
- It has the potential to address the current problem of pollution.
- It can provide for affordable, clean drinking water through swift detection of impurities and purification of water.
- The nanotechnology can be used to remove industrial water pollutants in the

groundwater through chemical reactions at a cheaper rate than the current methods that need pumping of the groundwater for treatment.

- Nanotechnology sensors and solutions also have the potential to detect, identify, filter and neutralize harmful chemical or biological agents in the air and soil.

Transport

- Nanotechnology contributes to manufacturing lighter, smarter, efficient and greener automobiles, aircraft and ships.

- It also allows various means to improve transportation infrastructures like providing resilience and longevity of the highway and other infrastructure components.

- The nanoscale sensors and devices can also provide for cheap and effective structural monitoring of the condition and performance of the bridges, rails, tunnels, etc. They can also enhance transportation infrastructure that makes the drivers avoid collisions and congestions, maintain lane position, etc.

Space

- Materials made of carbon nanotubes can reduce the weight of the spaceships while retaining or increasing the structural strength.

- They can also be used to make cables that are needed for the space elevator. Space elevators can significantly reduce the cost of sending materials to the orbit.

- The Nano sensors can be used to monitor the chemicals in the spacecraft to look into the performance of the life support system.

Agriculture

- The Nano capsule can enable effective penetration of herbicides, chemical fertilizers, and genes into the targeted part of the plant. This ensures a slow and constant release of the necessary substance to the plants with minimized environmental pollution.

- The Nano sensors and delivery systems can allow for precision farming through the efficient use of natural resources like water, nutrients, chemicals etc.

- The Nano sensors can also detect the plant viruses and soil nutrient levels.

- Nano-barcodes and Nano-processing could also be used to monitor the quality of agriculture produce.

Nanoscale

The macroscopic properties of any material (e.g., melting and boiling point, etc.) can be measured by examining the sample in conventionally convenient quantities under standard lab conditions. These quantities can vary for different materials as well as the purpose of investigations. One of the most often defined quantity is 'mole' such that one mole of any substance consists 6.022×10^{23} molecules. Thus, if any property of one mole of the substance is measure, this value usually represents the average value of 6.022×10^{23} molecules of that substance. It is often deduced that this value remains same regardless of the size of the group of molecules under investigation. However, this does not hold for all materials. That is, as the size of the material is decreased to reach nanoscale, the same material may demonstrate radically different properties. In other words, materials start demonstrating size dependent properties after reaching the nano-dimensions. This happens because matter at nanoscale does not follow Newtonian principles and quantum mechanics needs to be applied to describe the behavior of materials. What kind of small is this? Nanoscaled materials are considered to be consisting of clusters of atoms/molecules, not the single atoms/ molecules. For instance, 8 hydrogen atoms or 3.5 gold atoms placed adjacent forming a row become one nanometre long. In this regard, the size of nanomaterials is intermediate to a single atom or molecule, and their bulk forms. At these dimensions, matter exhibits novel properties. Therefore, nanomaterials offer the following opportunities:

- They can be used to improve existing materials as well as to produce new materials with exceptional properties.
- As nanomaterials have the size resembling that of the largest molecules present in nature (such as proteins, DNA, etc.), they can be integrated into a device to interact with these molecules.

Physics at Nanoscale

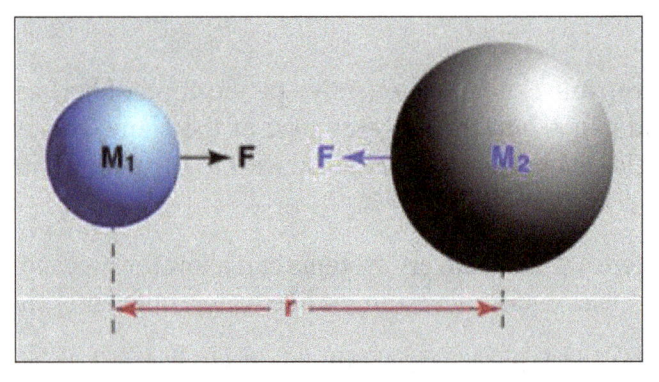

Owing to their nanoscaled dimensions, Nano materials are dimensionally more close to individual atoms and molecules than to the bulk materials, and therefore, their behavior is described using quantum mechanics. Quantum mechanics is the scientific model employed to describe the motion and energy of individual atoms and electrons. The significant quantum effects and the properties relevant at nanoscale are described below:

- At nanoscale, electromagnetic forces dominate whereas the gravitational forces are negligible.

Since the mass of nanoscaled objects becomes very small, gravitational force, which scales linearly with the mass of the particles, becomes negligible.

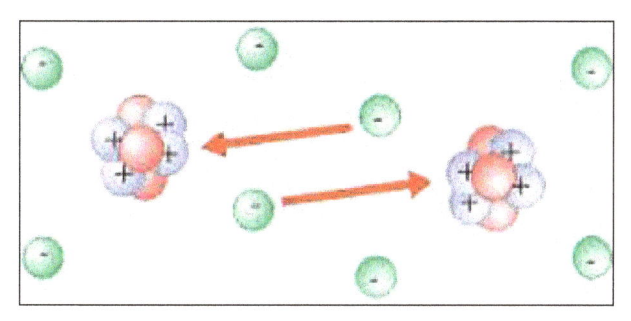

Since electromagnetic force is independent of the mass of the particles, and is determined by the charge and distance of the particles, it is strong even for the Nano sized particles. As an example, between two protons, electromagnetic force is 10^{36} times stronger than the gravitational force between them.

- Wave-particle duality: For extremely small objects having very low mass (e.g. electron), wave nature becomes pronounced. Thus an electron exhibits wave-like properties and its position can be described by the wave probability function.

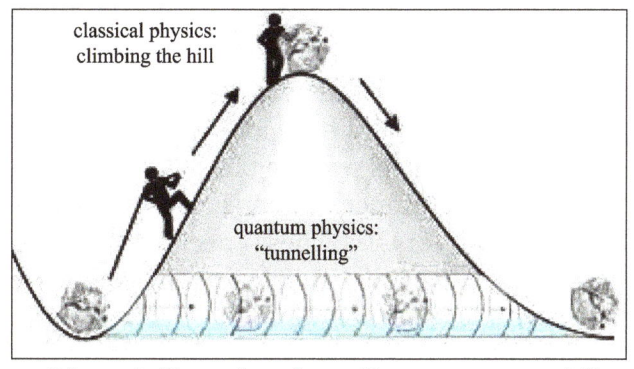

Figure: Schematic illustration of tunneling across a potential barrier.

- Tunneling: Tunneling is an extremely important consequence in nanoscaled materials. As per classical physics, an object can pass a potential barrier if its energy is greater than that of the barrier. Thus, if the object has energy lower than the barrier potential, the object cannot cross the barrier. In this case, there is zero probability of locating the object on other side of the potential barrier.

As per quantum physics, the object can tunnel through the barrier. Also, the thickness of the barrier is crucial in determining the tunneling probability of the particle. That is, the thickness of the barrier should be of the order of particle's wavelength. Thus, even if the object energy is lower than the potential of the barrier, there is a finite probability of locating the particle at the other side of the barrier.

Tunneling is an important quantum effect and forms the basic principle of scanning tunneling microscope (STM) used to image the nanostructured materials.

- Quantum Confinement: In nanostructured materials, including metals, electrons are not free to move within the material, but are confined in space.

- Quantization of energy: The electrons in the nanomaterials do not have continuous energy bands; rather they exist only at some discrete energy states. This is effect called the quantization of energy, and is most pronounced in quantum dots.

- Random molecular motion assumes importance: Above absolute zero, the molecules move owing to their kinetic energy. This motion is described as the random molecular motion and occurs in materials at a temperature above 0K. In bulk materials, this movement can be neglected in comparison to the size of the object; therefore do not influence object's movement. However, in nanoscaled objects, these motions are comparable to the particles' dimensions and thus affect their behavior.

- Larger surface area to volume ratio: Nanomaterials possesses very large surface areas. The smaller the object is, the larger the surface area to volume ratio. Large surface-to-volume ratio is an extremely important characteristic of a nanoparticle.

- Energy of confined electrons: In Nano-crystals, the energy levels of electrons do not remain continuous as is the case with bulk materials. Rather, owing to the confined electron wave functions, they become discrete resulting in finite DoS. These effects appear when the dimensions of the potential well approach de Broglie wavelength of electrons resulting in change in energy levels. The effect is described as Quantum confinement and consequently, the Nano-crystals are often called quantum dots or QDs. These effects influence the electrical, optical as well as mechanical behaviour of material. Depending on the QD size, confined electrons have higher energy than the electrons in bulk materials. This shift in energy is given by:

$$\Delta E = \frac{n^2 h^2}{8 m a^2}$$

Here, n represents the principal quantum number; h is Planck's constant; m is effective mass and a is the radius of the QD.

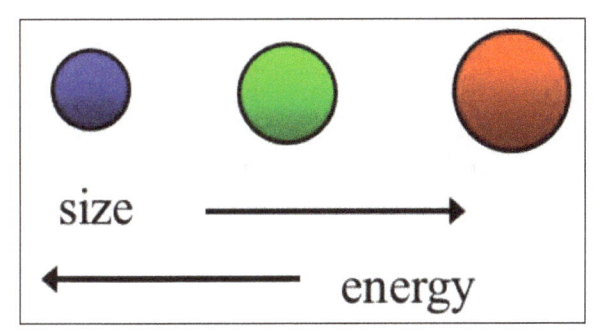

Figure: Variation of energy with size of the quantum dot (QD).

Chemistry at Nanoscale

Since a nanoparticle is essentially a cluster of atoms/molecules, all forms of bindings relevant in chemistry, are relevant at nanoscale as well. These are described below:

- Intramolecular bonding: These are also called chemical interactions and cause a change in the chemical structure of the molecule. Examples are ionic, metallic, and covalent bonding.

- Inter-molecular bonding: These are the physical interactions causing no change in chemical structure of molecule. Examples are ion-ion bonds, ion-dipole bonds, van der Waals bonds, hydrogen bonds, hydrophobic bonds, repulsive interactions, etc.

Nanomaterials are formed of a variety of molecules joined together or large molecules which take 3D structures via inter-molecular bonding (such as macromolecules). Thus Nanoscience includes supramolecular chemistry as well. Additionally, intermolecular bonding plays a significant role in these macromolecules.

- Intermolecular bondings include hydrogen and van der Waals interactions. Such bonds are individually weak. However, when they are present in large numbers, their total energy can be very large. As an example, a DNA molecule (cross-section = 2nm) has numerous hydrogen bonds holding the two helixes together. These interactions assume great significance in nanomaterials as their surface area is very large and numerous small forces apply on very large areas.

- Intermolecular bonds are essential to the structure of macromolecules (e.g., proteins). They form their specific 3D structure which has specific biological functions. Any disturbance to these bonds may permanently affect their structure which in turn may deteriorate their working.

- Hydrophobic effect is another significant intermolecular interaction in Nanoscience. This process is fundamentally driven by entropy. This property can be described as the ability of non-polar molecules (such as oil) to exist as clusters in water.

Molecules as "Devices"

Macromolecules, in Nanoscience, are frequently described as "devices". These devices can trap or release a certain species under specific conditions (e.g. pH). When molecules function as devices, bonds may also be device components. The application of molecules including molecular switches, actuators, and electronic wires is an important area of investigation.

Material Properties at Nanoscale

Surface Properties

Surface properties influence the physical and chemical properties of all the materials whether they are in bulk or nanoscale form. Surface performs several functions such as they manage inflow or outflow of materials or energy; they can work as catalysts, etc. The science which investigates the chemical, physical and biological properties of surface is termed as surface science. Generally, the surface is regarded as the interface, since it acts as a boundary between the material and its surroundings.

Nanomaterials possess large surface areas. This can be understood as follows: when the bulk material is repeatedly divided into smaller and smaller pieces, total volume, of all the pieces taken together, does not change while the surface area considerably increases. Thus, the surface-area to volume ratio is greatly enhanced.

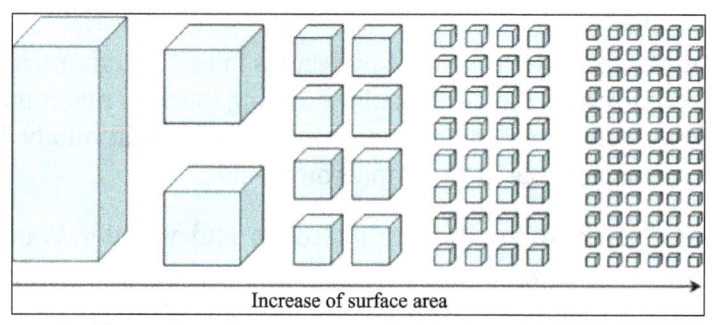

Figure: Schematics illustration of increase in surface-to-volume ratio with reduction in size.

Importance of Surface Atoms

The chemical groups present at the surface/interface of a material influence its properties (e.g., catalysis, electronic conduction, adhesion, gas storage, etc.). Since in nanomaterials, majority of atoms are present at their surface, they exhibit enhanced surface activities. This makes nanomaterials particularly useful in catalytic reactions, detection reactions including sensing of a specific compound, etc. which require physical adsorption of particular species at material's surface.

Additionally, properties like melting point are also affected by the size of the particle. Nano sized particles of a material have lower melting point than their bulk forms. This is because it is easy to remove the surface atoms as they are in contact with fewer atoms of the material, thus the energy required to overcome the inter-molecular forces holding the atoms is low leading to lower melting point. As an example, bulk CdSe melts at 1678K whereas a 3nm CdSe crystal melts at 700K.

Shape Matters

For a given volume, the surface area also depends upon material's shape, e.g., between a sphere and cube with equal volumes, the surface area of cube is larger. Consequently, apart from the size, shape of the material is also significant.

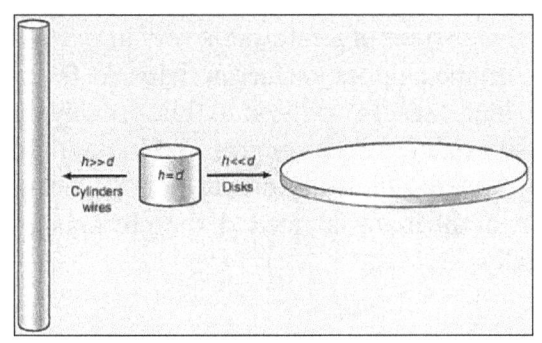

Figure: Shape affects the surface area.

Surface Energy Atoms or molecules existing at the interface (or surface) differ from the same atoms/molecules present in the interior or bulk of the material. This applies to all materials. Surface atoms or molecules have increased reactivities and are highly prone to form clusters. Thus, the surface atoms are unstable and possess higher surface energies.

The majority atoms of a nanomaterial are surface atoms. This causes high energies of the nanomaterials. Since, high energy systems strive to lose their energy, by any process; nanomaterials are inherently unstable. There are a variety of ways with which a nanomaterial can minimize its high surface energy. Agglomeration is a natural way to decrease the surface energy (as it strongly depends on the surface area, which can be greatly decreased via clustering). As shown in figure, by clubbing two boxes of similar surface areas, the overall surface area of the clubbed box is highly reduced, thereby also decreasing the surface energy.

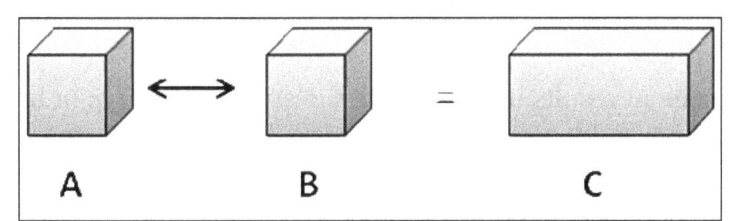

Figure: Surface energy of two isolated cubes is more than that of the two agglomerated cubes.

Thus, nanoparticles intrinsically tend to agglomerate. From their application point of view, agglomeration is undesirable. To reduce these tendency nanoparticles, a surfactant may be used.

Reactions where Surface Properties are Important

Catalysis

Catalysts are substances which enhance the rate of a chemical reaction without being consumed. Naturally occurring catalysts, often termed as enzymes, are highly efficient and produce the desired end products with minimal energy utilization. Artificial catalysts are usually produced by fixing metal particles over oxide surfaces. These are not energy efficient. The active surface of a catalyst is very important in catalyzing processes, since higher surface implies higher surface activity. As the size of the catalyst is reduced, its active surface increases. In addition to this, organization of active sites is also crucial. Since, both these features can be engineered in nanotechnology; it holds great potential to enhance catalyst design. Additionally, nanoparticles in catalysis will drastically reduce the quantity of the material needed, thereby benefitting both economically as well as environmentally.

Detection/Sensing

The detection of a specific chemical or biological species from a mixture is the basic principle of operation of sensing devices (e.g., chemical sensors, biosensors, and microarrays). Similar to catalysis, sensing is also a surface/interface phenomenon. The rate, specificity and accuracy of this reaction can be greatly enhanced by nanomaterials. Enhanced surface area in nanomaterials provides more active sites available for detection resulting in rapid detection. Also, the detection limit is also lowered. Further, as the nanomaterials with specific surface behaviors can be easily designed at molecular level, the active sites on the material surface can act as "locks" to detect specific molecules. Nanomaterials provide more detection sites in the same device, all of which can be engineered to detect a specific analyte/species. Therefore, nanotechnology can greatly enhance the detections capabilities in miniaturized devices.

Electrical Properties

Quantum Confinement and its Effect on Electrical Properties of Materials

Quantum confinement results in increase in the energy band gap of the material. Additionally, at nanoscale dimensions as the energy levels become quantized. Thus, the band overlap in metals disappears and a band gap emerges. Thus, some metals behave as semiconductors at nanoscale.

The increase in band gap energy implies that more energy can be absorbed by the material. Higher energies mean shorter wavelengths (or blue shift). A nanomaterial will emit the fluorescent light of higher wavelength resulting in blue shift. By adjusting the size of the nanoparticle, its absorption and emission can be tuned over a wide wavelength range.

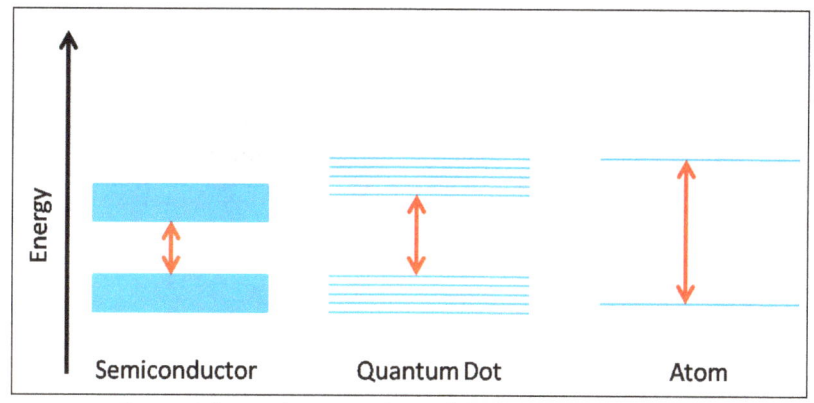

Figure: Schematic comparison of the band gap in the bulk semiconductor, a quantum dot and a single atom.

Optical Properties

The energy levels in nanoscaled semiconductors are quantized, splitting the conduction and valance bands. This discretizes the conduction as well as valance bands. Since charge can only be transferred across these discrete levels, only specific wavelengths can be absorbed. This results in monochromatic emissions from nanostructured semiconductors.

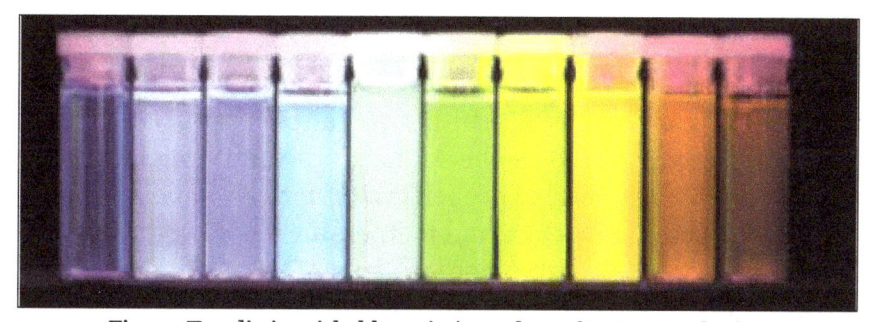

Figure: Ten distinguishable emission colors of ZnS-capped CdSe quantum dots excited by a near UV lamp.

Since the energy gap of the material is increased due to quantum confinement, more energy can be absorbed by it. Higher energy means shorter wavelength (blue shift). By tuning the size of the semiconductor Nano crystal, its band gap can be controlled. Thus, the wavelength absorbed/emitted light by the crystal can also be tuned. Consequently, same material (such as CdSe) can emit different colors depending on its size.

From White to Transparent Materials

High-protection sunscreens appear white due to scattering of visible light by them.

Most of them have ZnO and TiO$_2$ particles having dimensions of ~200 nm. When visible light interacts with these clusters, all of its wavelengths get scattered and combine to appear white colored.

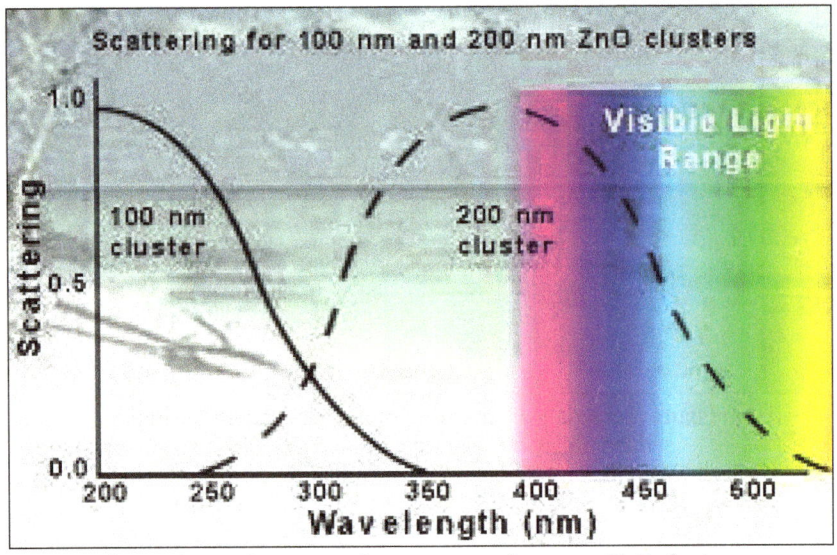

Figure: Scattering curves for 100 nm and 200 nm ZnO clusters.

When the dimension of the cluster is decreased down to ~100nm, maximum scattering occurs ~200nm, shifting the curve towards shorter wavelengths, which are not in visible range. Thus the same material appears transparent after reduction in size.

Magnetic Properties

Magnetic properties of the material can also be altered by Nano structuring. The magnetization curve of a material can be adjusted by controlling its size, resulting in soft or hard magnets with enhanced properties. Generally, the magnetic behavior of a material depends upon its structure and temperature. For experiencing a magnetic field, the material must have a finite nonzero spin. The transition size of classically expected magnetic domains is ~1μm. With decreasing the size of the magnet, surface atoms form a significant portion of the total number of atoms. This greatly enhances surface effects, and in turn the quantum confinement assumes significance. As the size of these domains is reduced to nanoscale, the materials exhibit novel properties because of quantum confinement, like the giant Magneto resistance effect (GME). This is a fundamental Nano-effect which is now being used in modern data storage devices.

Mechanical Properties

Depending on the structure, some nanomaterials can exhibit exceptional mechanical properties. One such material is carbon nanotubes (CNTs), which are extremely small tubes having the same honeycomb structure of graphite (bulk), but with different properties than it. CNTs can be 100 times stronger than steel but six times lighter.

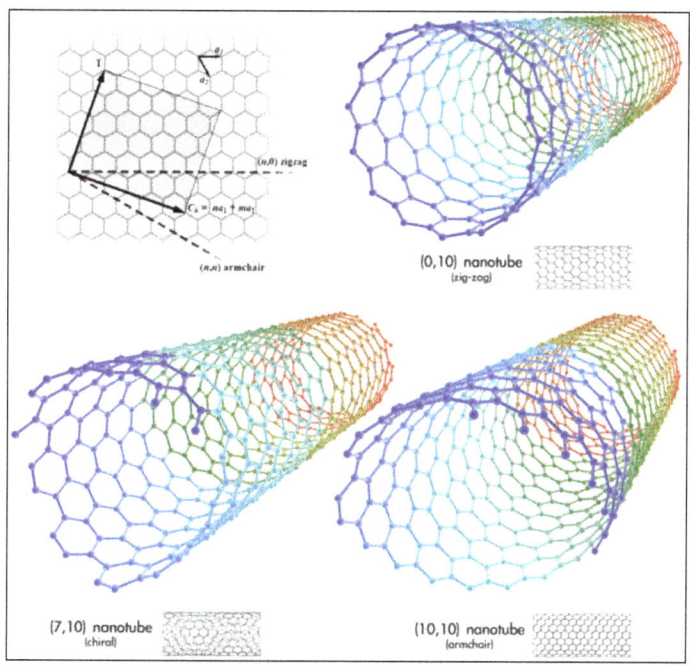

Figure: Various types of CNTs.

References

- Nanotechnology-india, Nanotechnology-is-the-development-and-one-billionth-of-a-meter: iasexpress.net, Retrieved 15, January 2020
- What-is-a-nanomaterial, resources, knowledgebase: safenano.org, Retrieved 29, August 2020

2

Nanoparticles: Synthesis and Application

The particles which have dimensions between 1 and 100 nanometers are called nanoparticles. The properties of nanoparticles are remarkably different from the larger particles of the same substance. Nanoparticles are often used to improve the properties of a material. This chapter has been carefully written to provide an easy understanding of the varied facets of nanoparticles.

Characterization of Nanoparticle

Morphological Characterization

The morphological features of NPs always attain great interest since morphology always influences most of the properties of the NPs. There are different characterization techniques for morphological studies, but microscopic techniques such as polarized optical microscopy (POM), SEM and TEM are the most important of these.

Figure: SEM images of ZnO modified MOFs at different temperatures.

SEM technique is based on electron scanning principle, and it provides all available information about the NPs at nanoscale level. Wide literature is available, where people used this technique to study not only the morphology of their nanomaterials, but also the dispersion of NPs in the bulk or matrix. The dispersion of SWNTs in the polymer matrix poly(butylene) terephthalate (PBT) and nylon-6 revealed through this technique. The same group also provides POM study of their materials, which showed star-like spherulites of the formed materials, whose size was decreased with the incremental filling of SWNTs. The morphological features of ZnO modified metal organic frameworks (MOFs) were studied through SEM technique, which indicates the ZnO NPs dispersion and morphologies of MOFs at different reaction conditions.

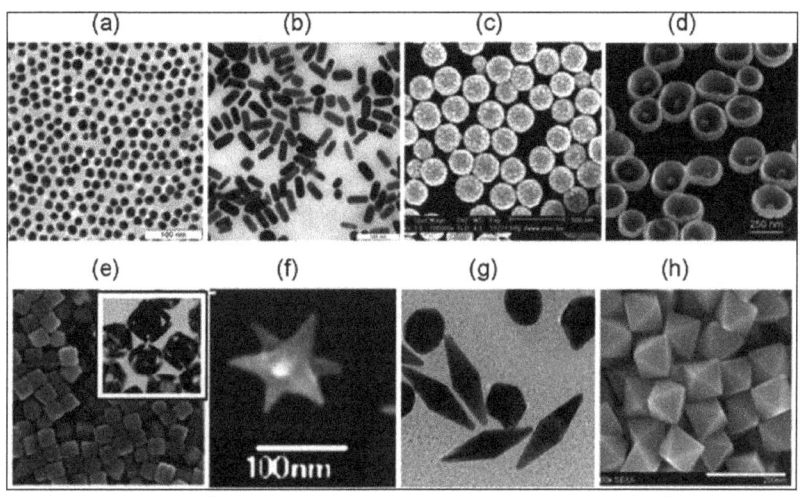

Figure: TEM images of different form of gold NPs, synthesized by different techniques.

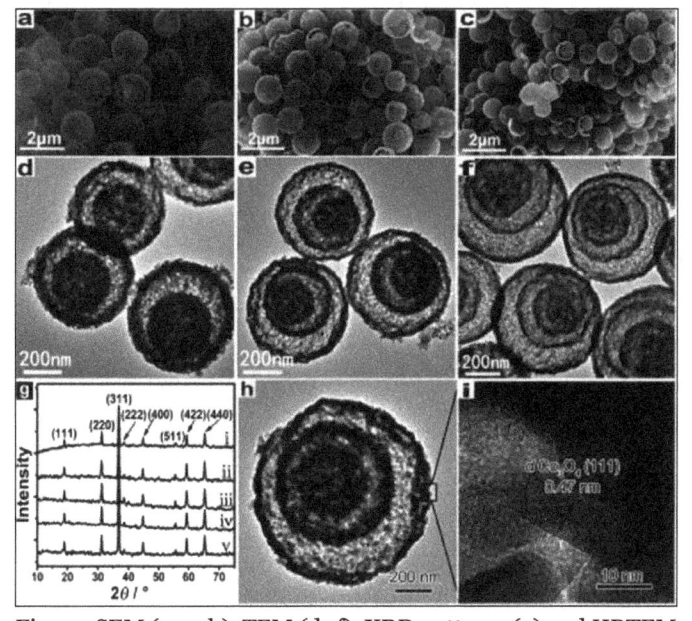

Figure: SEM (a–c, h), TEM (d–f), XRD patterns (g) and HRTEM (i) images of double, triple and quadruple Co_3O_4 hollow shells.

Similarly, TEM is based on electron transmittance principle, so it can provide information of the bulk material from very low to higher magnification. The different morphologies of gold NPs are studied via this technique. figure below provides some TEM micrographs showing various morphologies of gold NPs, prepared via different methods. TEM also provides essential information about two or more layer materials, such as the quadrupolar hollow shell structure of Co_3O_4 NPs observed through TEM. These NPs founded to be exceptionally active as anode in Li^- ion batteries. Porous multishell structure induces shorter Li^+ diffusion path length with adequate annulled space to buffer the volume expansion, good cycling performance, greater rate capacity, and specific capacity as well.

Structural Characterization

The structural characteristics are of the primary importance to study the composition and nature of bonding materials. It provides diverse information about the bulk properties of the subject material. XRD, energy dispersive X-ray (EDX), XPS, IR, Raman, BET, and Zieta size analyzer are the common techniques used to study structural properties of NPs.

XRD is one of the most important characterization techniques to reveal the structural properties of NPs. It gives enough information about the crystallinity and phase of NPs. It also provides rough idea about the particle size through Debye Scherer formula. This technique worked well in both single and multiphase NPs identification. Nevertheless, in the case of smaller NPs having size less than hundreds of atoms, the acquisition and correct measurement of structural and other parameters may be difficult. Moreover, NPs having more amorphous characteristics with varied inter atomic lengths can influence the XRD diffractogram. In that case, proper comparison of the diffractograms of bimetallic NPs with those of the corresponding monometallic NPs and their physical mixtures is required to obtain accurate information. Comparison of computer simulated structural model of bimetallic NPs with observed XRD spectra is the best way to get good contrast. EDX, which is normally fixed with field emission scanning electron miscopy (FE-SEM) or TEM device, is widely used to know about the elemental composition with a rough idea of % wt. The electron beam focused over a single NP by SEM or TEM through the program functions, to acquire the insight information from the NP under observation. NP comprises of constituent elements and each of them emits characteristics energy X-rays by electron beam irradiation. The intensity of specific X-ray is directly proportional to the concentration of the explicit element in the particle. This technique is widely used by researchers to give support to SEM and other techniques for the confirmation of their elements in prepared materials. The EDX technique used to determine the elemental composition of ultra-sonochemically synthesized pseudo-flower shaped $BiVO_4$ NPs. Similarly, by utilizing similar technique the elemental confirmation and graphene impregnation of In_2O_3/graphene heterostructure NPs was carried out, which showed C, In and O as contributing elements. This material was synthesized through conventional hydrothermal technique.

XPS is considered to be the most sensitive technique and it is widely used to determine the exact elemental ratio and exact bonding nature of the elements in NPs materials. It is surface sensitive technique and can be used in depth profiling studies to know the overall composition and the compositional variation with depth. XPS is based on the basic spectroscopic principles and typical XPS spectrum is composed of the number of electrons on Y-axis plot versus the binding energy (eV) of the electrons on X-axis. Each element has their own fingerprint binding energy value and thus gives specific set of XPS peaks. The peaks correspond come from electronic configuration, e.g., 1s, 2s, 2p, and 3s. Lykhach et al. provide a depth electron transfer study through CeO_2 supported Pt NPs using XPS technique with support to others. They determined that per ten Pt atoms, only one electron is elated from the NPs to CeO_2 support. The depth profile analysis was provided to study the dispersion of boron NPs (10 nm size) during polyethylene glycol (PEG) functionalization. Ar^+ ions were used at 1.4 keV and 20 nm; depth surface etching was performed. It was revealed that the concentration of NPs increases from 2 to 5% with depth. This provided good evidence that boron NPs are dissolved effectively within the bulk of functionalized PEG. In similar study core shell Au/Ag showed similar behavior through XPS depth profiling. Wang et al. quantify the NPs coating with this technique through XPS and STEM spectroscopies by help of SESSA software.

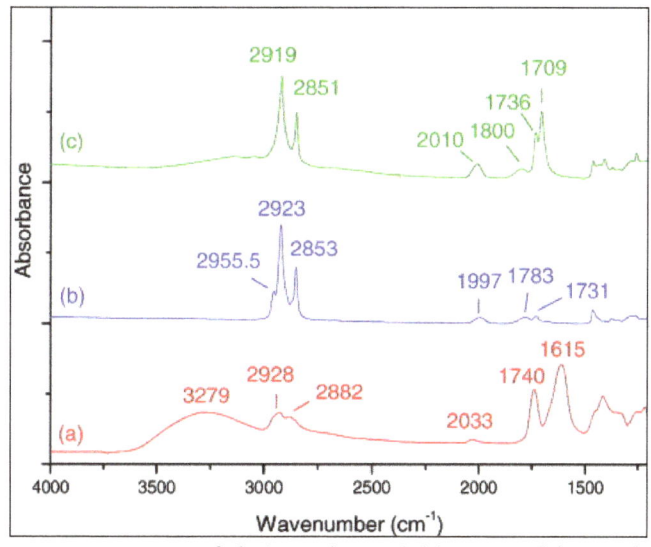

Figure: FTIR spectra of platinum (1.7 nm) (a) extracted from polyol, (b) dodecanethiol coated Pt, and (c) MUDA coated Pt.

Vibrational characterization of nanoparticles is normally studied via FT-IR and Raman spectroscopies. These techniques are the most developed and feasible as compared to other elemental analytical methods. The most important range for NPs is the fingerprint region, which provides signature information about the material. In one study, functionalization of Pt NPs (1.7 nm mean size) and its interaction with Alumina substrate studied via FT-IR and XPS technique. FT-IR confirms the functionalization as it showed the signature vibrational peaks of carboxylated C–O 2033 cm^{-1}, respectively in addition to a broader O–H peak at 3280 cm^{-1}. The degree of functionalization was revealed from the red shift values of FT-IR bands.

In another study, a series of 5 mol% Eu^{3+} – doped rare earth metal (RE) hafnium oxide $RE_2Hf_2O_7$ (where RE = Y, Pr, La, Gd, Lu and Er) (NPs) was synthesized by correlated techniques. FT-IR and Raman spectra analysis exhibited that the $La_2Hf_2O_7$:5%Eu^{3+} and $Pr_2Hf_2O_7$:5%Eu^{3+} possessed relatively ordered pyrochlore structure as compared to RE$_2$Hf$_2$O$_7$:5%Eu^{3+} compositions (RE = Y, Er, and Lu), which possess disordered fluorite structure. The stable structures were found thermodynamically stable until high temperature of 1500 °C. However, disordered–ordered cause instability in the latter case, and thus it is thermodynamically unstable.

More recently surface enhanced Raman spectroscopy (SERS) is evolving as vibrational conformational tool due to its signal enhanced capability via SPR phenomenon. One study reported SERS technique to study the vibrational properties with phonons modes in nanostructured and quantum dots NPS of TiO_2, ZnO and PbS. They concluded that the enhanced spectra can be attributed to the plasmonic resonances in semiconductor systems.

Optical Characterization

Optical properties are of great concerned in photocatalytic applications and therefore, photo-chemists acquired good knowledge of this technique to reveal the mechanism of their photochemical processes. These characterizations are based on the famous beer-lambert law and basic light principles. These techniques give information about the absorption, reflectance, luminescence and phosphorescence properties of NPs. It is widely known that NPs especially metallic and semiconductor NPs possess different colors and therefore, best harmonized for photo-related applications. So, it is always interesting to know the value of absorption and reflectance of these materials to understand the basic mechanism for each application. Ultraviolet–visible (UV–Vis), photoluminescence (PL) and the null ellipsometer are the well-known optical instruments, which can be used to study the optical properties of NPs materials.

The UV/vis diffuse reflectance spectrometer (DRS) is a fully equipped device which can be used to measure the optical absorption, transmittance and reflectance. The former two are supplementary to each other while the latter (DRS) is a special technique use for sold samples mostly. The method is exceptionally acceptable for the determination of band gaps of NPs and other nanomaterials. Band gap of materials is very important to conclude about the photoactivity and conductance of the material. The carbon nanodot-carbon nitride (C_3N_4) was found to be a metal free water splitting photo catalyst. The photo ability of this material is directly correlated to the band gap value of 2.74–2.77 eV, which was calculated using UV–Vis spectroscopy. Similarly, this technique also uses to see the absorption shift in case of doping, composite formation or heterostructure NPs materials. Peng et al. synthesis MMT, LaFeO$_3$ and LaFeO$_3$/MMT Nano composites and studied variation in their electromagnetic radiations absorption through UV–vis DRS to reconnoiter their optical characteristics. The strong red shift observed in case of Nano composite as compared to pristine MMT and LaFeO$_3$ NPs. LaFeO$_3$ and LaFeO$_3$/MMT displayed rather broad absorption band from 400 to 620 nm, showing decrease in their band gap. This property makes these catalysts considerable for solar light driven photocatalysis.

In addition to UV, PL also considers valuable technique to study the optical properties of the photoactive NPs and other nanomaterials. This technique offers additional information about the absorption or emission capacity of the materials and their effect on the overall excitation time of photoexcitons. Thus, it provides significant information about the charge recombination and half-life of the excited materials in their conductance band, which are useful for all photo related and imaging applications. The PL spectrum can be recorded as emission or absorbance depending on the nature of study. Figure below shows a typical PL spectrum of pristine and modified ZnO NPs. It is evident from this figure that pristine ZnO NPs show high PL intensity as compared to CdS modified ZnO NPs. The gold embedded CdS/Au/ZnO composite shows the lowest intensity. This quenching from pure ZnO to CdS/Au/ZnO can be attributed to the decrease in the rate of charge recombination and larger lifetime of photoexcitons in the latter case. In addition, this technique is successfully used to determine the thickness of layer, doping quantity of material and defects/oxygen vacancies determination of NPs.

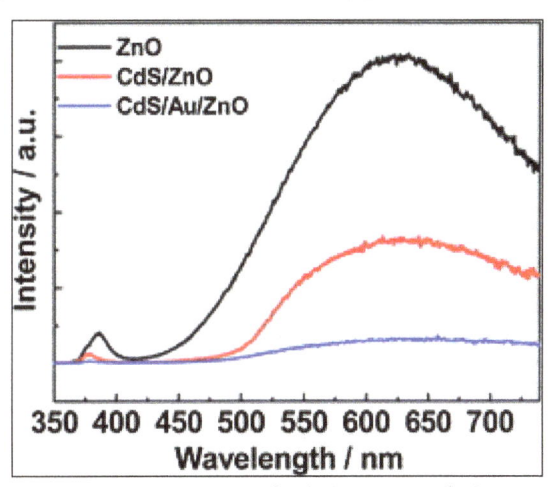

Figure: Photoluminescence (PL) spectra of pristine ZnO, CdS/ZnO, and CdS/Au/ZnO measured with 270 nm excitation wavelength at normal temperature.

Similarly, researchers determined the values of refractive index and extinction coefficient for hollow gold NPs (HG-NPs) via spectroscopic ellipsometry. They prepared a series of HG-NPs, with different morphologies and plasmonic properties and the optical constants were calculated. The values were compared with the optical constant values of solid gold NPs, which gave good indication to use these materials in chemical sensing applications due to their sensitive nature as revealed from ellipsometric values.

Structure of Nanoparticles

Crystal Structure

Many properties of solids depend on the size range over which they are measured. Microscopic details become averaged when studying bulk materials. At the macro or large scale range ordinarily studied in traditional fields of physics such as mechanics, electricity and magnetism and optics, the sizes of the objects under study range from millimeter to kilometers. The properties that we associate with these materials are averaged properties, such as the density and elastic moduli in mechanics, the resistivity and magnetization in electricity and magnetism, and the dielectric constant in optics. However, when the properties are measured in micrometer and nanometer range, many properties of materials change, such as mechanical and ferroelectric, and ferromagnetic properties. Below the nanometer range, there is the atomic scale near 0.1 nm, followed by the nuclear scale near a Fermi.

In its solid form materials are characterized into two categories: a) Crystalline and b) amorphous. Crystalline materials have so called long range ordering as regularity can extend throughout the crystal while amorphous materials are short ranges. Examples of the crystalline materials are NaCl, KCl, etc. and glass, wax etc. are the examples of amorphous materials. Gases lack both long range and short range order.

A two dimensional crystal have five type lattice ordering a) square; b) primitive rectangular; c) centered rectangular; d) hexagonal and e) oblique types, which is shown in the following figure below.

square	$a_1 = a_2$	$\gamma = 90°$
hexagonal	$a_1 = a_2$	$\gamma = 120°$
rectangular	$a_1 = a_2$	$\gamma = 90°$
centered rectangular	$a_1 = a_2$	$\gamma = 90°$
oblique	$a_1 = a_2$	$\gamma \neq 90°, 120°$

Figure: The five Bravais lattices that occur in two dimensions, with the unit cells indicated: (a) square; (b) hexagonal; (c) primitive rectangular; (d) centered rectangular; (e) oblique.

These arrangements are called Bravais lattices. The general or oblique Bravais lattice has two unequal lattice constants a ≠ b and an arbitrary angle θ between them. For the perpendicular case when θ = 90°, the lattice becomes the rectangular type. For the special case a = b and θ = 60°, the lattice is the hexagonal type formed equilateral triangles. Each lattice has a unit cell, indicated in the figures, which can replicate throughout the plane and generate the lattice.

In three dimensions there are three lattice constants, a, b and c and three angles: α between b and c; β between c and a; and γ between lattice constants a and b. There are 14 Bravais lattices, ranging from the lower symmetry triclinic type in which all three lattice constants and all three angles differ from each other (a ≠ b ≠ c and α ≠ β ≠ γ), to the highest symmetry cubic case in which all the lattice constants are equal and all the angles are 90° (a = b = c and α = β = γ = 90°). There are three Bravais lattices in the cubic system, namely, a primitive or simple cubic (SC) lattice in which the atoms occupy the eight corners of the cubic unit cell, a body centered cubic (BCC) lattice points occupied at the corners and in the center of the unit cell and a face centered cubic (FCC) Bravais lattice with atoms at the corners and in the centers of the faces.

Types of Bravais Lattice

Out of 14 types of Bravais lattices some 7 types of Bravais lattices in three-dimensional space are as follows. Note that the letters a, b, and c have been used to denote the dimensions of the unit cells whereas the letters α, β, and γ denote the corresponding angles in the unit cells.

Cubic Systems

In Bravais lattices with cubic systems, the following relationships can be observed.

a = b = c

α = β = γ = 90°

The 3 possible types of cubic cells have been illustrated below:

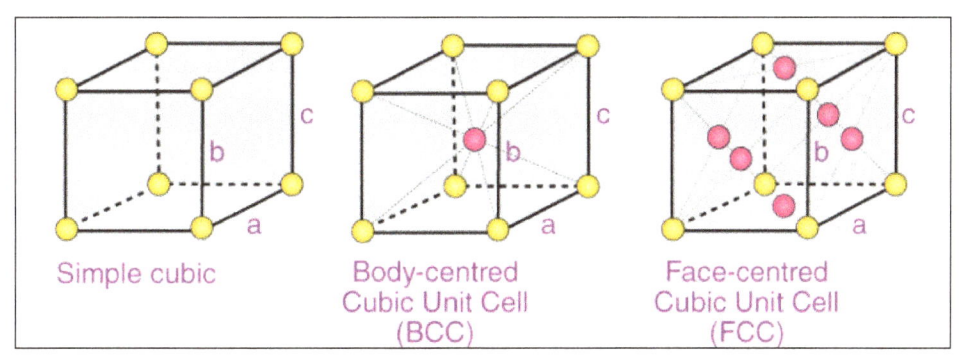

These three possible cubic Bravais lattices are:

- Primitive (or Simple) Cubic Cell (P).
- Body-Centered Cubic Cell (I).
- Face-Centered Cubic Cell (F).

Examples: Polonium has a simple cubic structure, iron has a body-centered cubic structure, and copper has a face-centered cubic structure.

Orthorhombic Systems

The Bravais lattices with orthorhombic systems obey the following equations:

$a \neq b \neq c$

$\alpha = \beta = \gamma = 90°$

The four types of orthorhombic systems (simple, base centered, face-centered, and body-centered orthorhombic cells) are illustrated below:

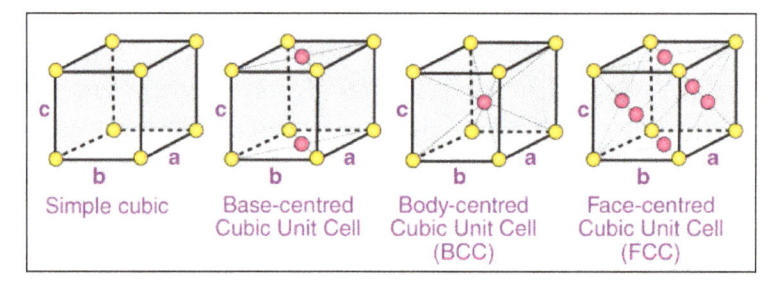

Examples of Orthorhombic Systems:

- Rhombic sulphur has a simple orthorhombic structure.
- Magnesium sulfate heptahydrate ($MgSO_4 \cdot 7H_2O$) is made up of a base centered orthorhombic structure.
- Potassium nitrate has a structure which is body-centered orthorhombic.
- An example of a substance with a face-centered orthorhombic structure is barium sulfate.

Tetragonal Systems

In tetragonal Bravais lattices, the following relations are observed:

$a = b \neq c$

$\alpha = \beta = \gamma = 90°$

The two types of tetragonal systems are simple tetragonal cells and body-centered tetragonal cells, as illustrated below:

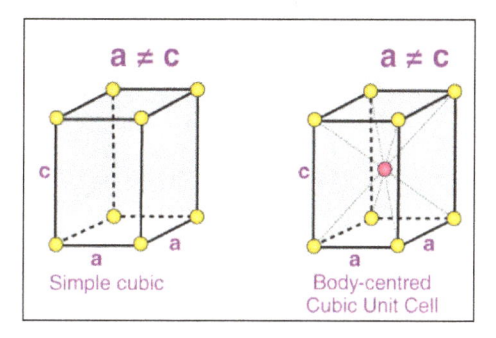

Examples of tetragonal Bravais lattices are – stannic oxide (simple tetragonal) and titanium dioxide (body-centered tetragonal).

Monoclinic Systems

Bravais lattices having monoclinic systems obey the following relations:

$a \neq b \neq c$

$\beta = \gamma = 90°$ and $\alpha \neq 90°$

The two possible types of monoclinic systems are primitive and base centered monoclinic cells, as illustrated below:

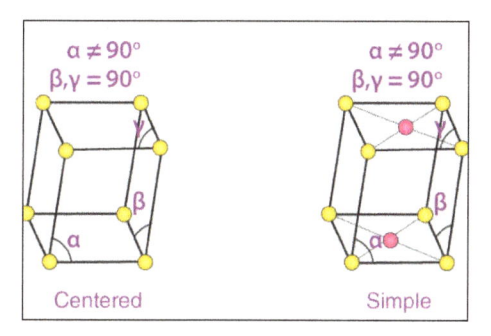

Cubic cells are – Monoclinic sulphur (simple monoclinic) and sodium sulfate decahydrate (base centered monoclinic).

Triclinic System

There exists only one type of triclinic Bravais lattice, which is a primitive cell. It obeys the following relationship.

$a \neq b \neq c$

$\alpha \neq \beta \neq \gamma \neq 90°$

An illustration of a simple triclinic cell is given below:

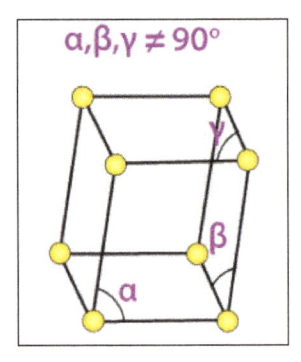

Such unit cells are found in the structure of potassium dichromate (Chemical formula $K_2Cr_2O_7$).

Rhombohedral System

Only the primitive unit cell for a rhombohedral system exists. Its cell relation is given by:

$a = b = c$

$\alpha = \beta = \gamma \neq 90°$

An illustration of the primitive rhombohedral cell is provided below:

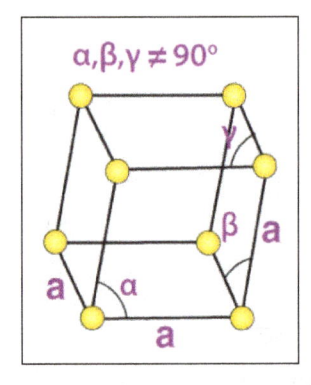

Calcite and sodium nitrate are made up of simple rhombohedral unit cells.

Hexagonal System

The only type of hexagonal Bravais lattice is the simple hexagonal cell. It has the following relations between cell sides and angles.

$a = b \neq c$

$\alpha = \beta = 90°$ and $\gamma = 120°$

An illustration of a simple hexagonal cell is provided below:

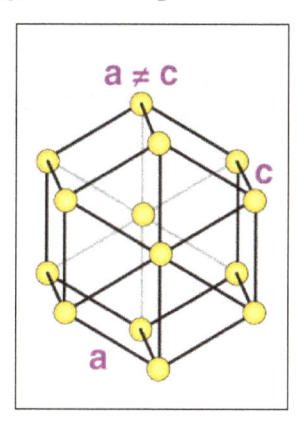

Zinc oxide and beryllium oxide are made up of simple hexagonal unit cells. Thus, it can be noted that all 14 possible Bravais lattices differ in their cell length and angle relationships.

2-Dimensional Structures

In two dimensions the most efficient way to pack identical circles (or spheres) is the equilateral triangle arrangement, corresponding to the hexagonal Bravais lattice. A second hexagonal layer of spheres can be placed on top of the first to form the most efficient packing of two layers. For efficient packing, the third layer can be placed either above the first layer with an atom at the location indicated by T or in the third possible arrangement with an atom above the position marked by X. In the first case a hexagonal lattice with a hexagonal close packed (HCP) structure is generated, and in the second case, a face centered cubic lattice results. In the three dimensional case of close packed spheres there are spaces or sites between the spheres where smaller atoms can reside. The point marked by X on figure, called an octahedral site, is equidistant from the three spheres O below it and from the three spheres O above it. An atom A at this site has the local coordination AO_6. The radius a_{oct}, of this octahedral site is:

$$a_{oct} = \frac{1}{4}\left(2-\sqrt{2}\right) a = \left(\sqrt{2}-1\right) a_o = 0.41421_{ao}$$

where a is the lattice constants and ao is the radius of the spheres. The number of octahedral sites is equal to the number of spheres. There are also smaller sites, called tetrahedral sites, labeled T in the figure that are equidistant from the nearest neighbor spheres, one below and three above, corresponding to AO_4 for the local coordination. This is a smaller site since its radius a_T is:

$$a_T = \frac{1}{4}\left(\sqrt{3}-\sqrt{2}\right) a = \left[\left(\frac{3}{2}\right)^{1/2} - 1\right] a_o = 0.2247_{ao}$$

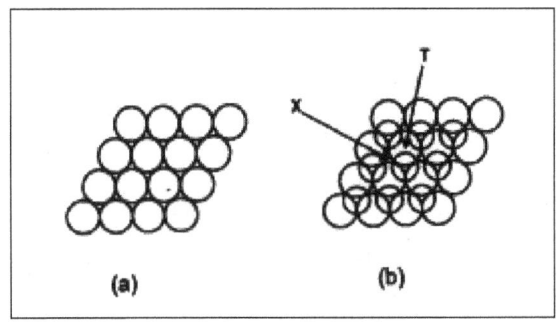

Figure: Close packing of spheres on a flat surface: (a) for a monolayer; (b) with a second layer added. The circles of the second layer are drawn smaller for clarity. The location of an octahedral site is indicated by X and the position of a tetrahedral site is designated by T on panel (b).

There are twice as many tetrahedral sites as there are spheres in the structures. Many diatomic oxides and sulfides such as MgO, MgS, MnO and MnS have their larger oxygen or sulfur anions in a perfect FCC arrangement with the smaller metal cations located at octahedral sites. This is called NaCl lattice type, where we use the term anion for a negative ion and cation for a positive ion. The mineral spinel $MgAl_2O_4$ has a face centered arrangement of divalent oxygen O^{2-} (radius 0.132 nm) with the Al^{3+} ions (radius 0.051 nm) occupying one half of the octahedral sites and Mg^{2+} (radius 0.066 nm) located in one-eighth of the tetrahedral sites in a regular manner.

Face Centered Cubic Nanoparticles

Most metals in the solid state form close packed lattices; thus Ag, Al, Au, Cu, Co, Pb, Pt and Rh, as well as rare gases Ne, Kr and Xe are face centered cubic (FCC), and Mg, Nd, Os, Re, Ru, Y, and Zn are hexagonal close packed (HCP). A number of other metallic atoms crystallize in the not so closely packed body-centered cubic (BCC) lattice, and a few such as Cr, Li and Sr crystallize in all three structure types, depending on the temperature. An atom in each of the two close packed lattices has 12 nearest neighbors. Figure below shows the 12 neighbors that surround an atom (darkened circle) located in the center of a cube for a FCC lattice. It also shows the 14 sided polyhedron, called a dekatessarahedron that is generated by connecting the atoms with planar faces. Sugano and Koizumi call this polyhedron a cuboctahedron. The three open circles at the upper right of figure are the three atoms in the top layer. The six, darkened circles plus an atom in the center of the cube of figure constitute the middle layer and the open circle at the lower left of figure is one of the three obscured atoms in the plane below the cluster. This 14 sided polyhedron has six square faces and eight equilateral triangles faces.

If another layer of 42 atoms is laid down around the 13 atom nanoparticles, one obtains a 55 atom nanoparticle with the same dekatessarahedron shape. Larger nanoparticles with the same polyhedron shape are obtained by adding more layers, and the sequence of numbers on the resulting particles, N = 1, 13, 55, 147, 309, 561 which are listed in table below, are called structural magic numbers.

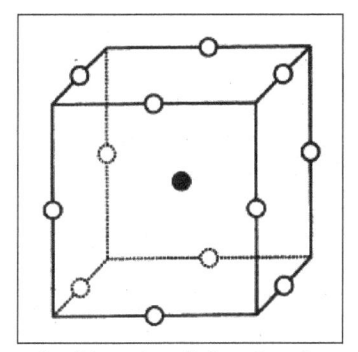

Figure: Face centered cubic unit cell showing the 12 nearest neighbor atoms that surround the atom (darkened circle) in the center.

For n layers the numbers of atoms N in these FCC nanoparticles is given by the formula:

$$N = \frac{1}{3}\left[10n^3 - 15n^2 - 11n - 3\right]$$

and the number of atoms in the surface N_{surf} is:

$$N_{surf} = 10n^2 - 20n + 12$$

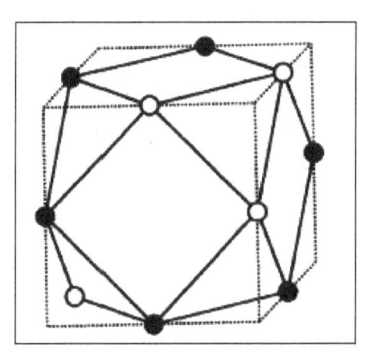

Figure: Thirteen atom nanoparticle set in its FCC unit cell, showing the shape of the 14 sided polyhedron associated with the Nano cluster.

The three open circles at the upper right correspond to the atoms of the top layer, the six solid circles plus the atom (not pictured) in the center of the cube constitute the middle hexagonal layer and the open circles at the lower left corner of the cube is ine of the three atoms at the bottom of the cluster.

For each value of n, table below lists the numbers on the surface, as well as the percentage of atoms on the surface. The table also lists the diameter of each nanoparticle, which is given by the expression (2n - 1) d, where d is the distance between the corners of nearest neighbor atoms and $d = \frac{a}{\sqrt{2}}$, where a is the lattice constant. If the same procedure is used to construct nanoparticles with the hexagonal close packed structure that was discussed in the previous paragraph, a slightly different set of structural magic numbers is obtained, namely, 1, 13. 57, 153, 321, 581.

Table: Number of atoms (structural magic numbers) in rare gas or metallic nanoparticles with face centered cubic close packed structures.

Shell Number		Number of FCC Nanoparticle Atoms		
	Diameter	Total	On surface	% Surface
1	1d	1	1	100
2	3d	13	12	92.3
3	5d	55	42	76.4
4	7d	147	92	62.4
5	9d	309	162	52.4
6	11d	561	252	44.9
7	13d	923	362	39.2
8	15d	1415	492	34.8
9	17d	2057	642	31.2
10	19d	2869	812	28.3
11	21d	3871	1002	25.9
12	23d	5083	1212	23.8
25	49d	4.9×10^4	5.76×10^3	11.7
50	99d	4.04×10^5	2.4×10^4	5.9
75	149d	1.38×10^6	5.48×10^4	4.0
100	199d	3.28×10^6	9.8×10^4	3.0

The diameters d in nanometers for some representative FCC atoms are Al 0.286, Ar 0.376, Au 0.288, Cu 0.256, Fe 0.248, Kr 0.400, Pb 0.350 and Pd 0.275.

Purely metallic FCC nanoparticles such as Au_{55} tend to be very reactive and have short lifetimes. They can be ligand-stabilized by adding atomic groups between their atoms and on their surfaces. The Au_{55} nanoparticles has been studied in the ligand-stabilized from $Au_{55}(PPh_3)_{12}C_{16}$ which has the diameter of ~1.4 nm, where PPh_3 is an organic group. Further examples are the magic numbers nanoparticles Pt_{309}(1, 10-phenantroline)$36O30$ and $Pd561(1, 10\text{-phenantroline})_{36}O_{200}$.

The magic numbers that we have been discussing are called structural magic numbers because they arise from minimum-volume, maximum-density Nano-particles that approximate a spherical shape, and have close packed structures characteristics of a bulk solid. These magic numbers take no account of the electronic structure of the constitute atoms in the nanoparticles. Sometimes the dominant factor in determining the minimum-energy structure of small nanoparticles is the interactions of the valence electrons of the constituents' atoms with this potential, so that the electrons occupy orbital levels associated with this potential. Atomic cluster configurations in which these electrons fill closed shells are especially stable, and constitute electronic magic numbers.

When mass spectra were recorded for sodium nanoparticles Na_N, it was found that mass peaks corresponding to the first 15 electronic magic numbers N = 3, 9, 20, 36, 61 were observed for cluster sizes up to N = 1220 atoms (n = 15), and FCC structural magic numbers starting with N = 1415 for n = 8 were observed for larger sizes. The mass spectral data versus the cube root of the number of atoms $N^{1/3}$ are plotted in figure and it is clear that the lines from both sets of magic numbers are approximately equally spaced, with the spacing between the structural magic numbers about 2.6 times that between the electronic ones. This result provides evidence that small clusters tend to satisfy electronic criteria and large structures tend to be structurally determined.

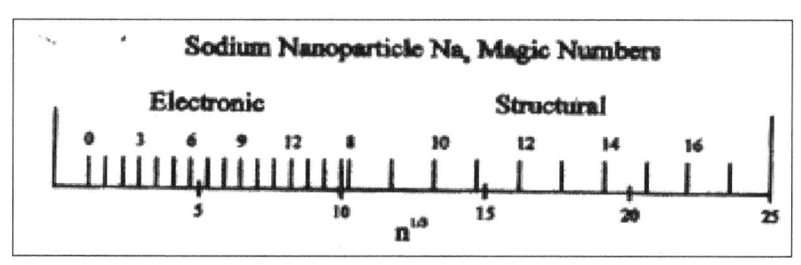

Figure: Dependence of the observed mass spectra lines from NaN, nanoparticles on the cube root N1/3 of the number of atoms N in the cluster. The lines are labeled with the index n of their electronic and structural magic numbers obtained from Martin et al.

Energy Bands

Formation of Energy Bands

In an isolated atom, the electrons in each orbit possess definite energy. But, in the case of solids, the energy level of the outermost orbit electrons is affected by the neighboring atoms.

When two isolated charges are brought close to each other, the electrons in the outermost orbit experience an attractive force from the nearest or neighboring atomic nucleus. Due to this reason, the energies of the electrons will not be at the same level, the energy levels of electrons are changed to a value which is higher or lower than that of the original energy level of the electron.

The electrons in the same orbit exhibit different energy levels. The grouping of this different energy levels is known as energy band. However, the energy of the inner orbit electrons are not much affected by the presence of neighboring atoms.

Classification of Energy Bands

Valence Band

The electrons in the outermost shell are known as valence electrons. These valence electrons contain a series of energy levels and form an energy band known as valence band. The valence band has the highest occupied energy.

Conduction Band

The valence electrons are not tightly held to the nucleus due to which a few of these valence electrons leave the outermost orbit even at room temperature and become free electrons. The free electrons conduct current in conductors and are therefore known as conduction electrons. The conduction band is one that contains conduction electrons and has the lowest occupied energy levels.

Forbidden Energy Gap

The gap between the valence band and the conduction band is referred to as forbidden gap. As the name suggests, the forbidden gap doesn't have any energy and no electrons stay in this band. If the forbidden energy gap is greater, then the valence band electrons are tightly bound or firmly attached to the nucleus. We require some amount of external energy that is equal to the forbidden energy gap. The figure below shows the conduction band, valence band and the forbidden energy gap.

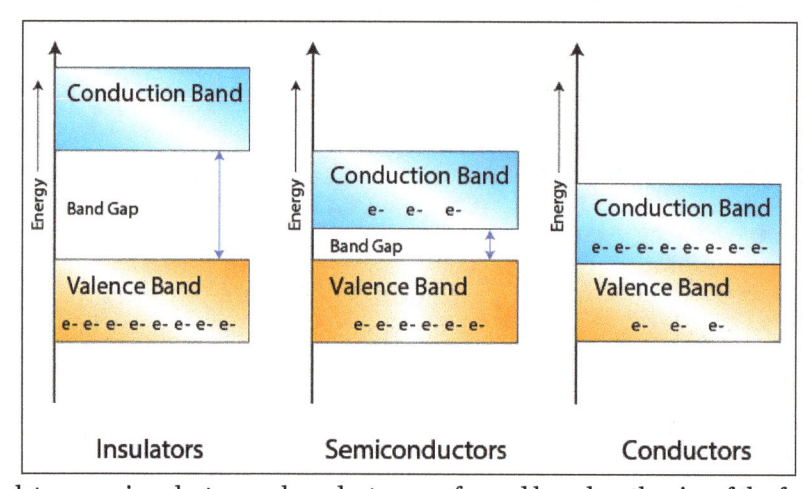

Figure: Insulators, semiconductors and conductors are formed based on the size of the forbidden gap.

Conductors

Gold, Aluminum, Silver, Copper, all these metals allow an electric current to flow through them. There is no forbidden gap between the valence band and conduction band which results in the overlapping of both the bands. The number of free electrons available at room temperature is large.

Insulators

Glass and wood are examples of the insulator. These substances do not allow electricity to pass through them. They have high resistivity and very low conductivity. The energy gap in the insulator is very high up to 7eV. The material cannot conduct because the movement of the electrons from the valence band to the conduction band is not possible.

Physicochemical Properties

Melting Point

Melting temperature relates to cohesive energy refers of the materials. It is the energy required to divide the metallic crystal into individual atoms. It also refers to heat of sublimation that can be determined experimentally or can be calculated using cellular method and density function theory. All these methods will calculate only for bulk material. The properties of nanoparticle vary due to the size effect. A simple method to calculate the cohesive energy of nanoparticle was discussed below. The cohesive energy increases with the increase in the particle size. When the particle size is large the cohesive energy will approach bulk material. Let a metallic particle has a diameter of D and is composed of n atoms. The surface area So of the particle given by,

$$S_o = \pi D^2$$

Assuming when the particle is separated into n identical spherical atoms and let the diameter of the atoms are d without changing its volume by exerting energy E_n we can write,

$$\frac{4}{3}\pi\left(\frac{D}{2}\right)^3 = n\frac{4}{3}\pi\left(\frac{d}{2}\right)^3$$

where,

$$n = \frac{D^3}{d^3}$$

The surface area of n atoms is,

$$S = n\pi d^2$$

When the particle is changed to n atoms the surface area variation is,

$$\Delta S = n\pi d^2 \pi D^2$$

Let E_n be the cohesive energy of n atoms and equals to the surface energy of the solid whose surface area is ΔS. The surface energy per unit area at oK is γ_o then $E_n = \Delta S\, \gamma_o$. (i.e.) $E_n = \pi\gamma_o(nd^2 - D^2)$.

Then the cohesive energy per atom is,

$$E_n = \pi\gamma_o\left(d^2 - \frac{D^2}{n}\right)$$

From equation, $n = \dfrac{D^3}{d^3}$

$$E = \pi\gamma_o d^2\left(1 - \frac{d}{D}\right)$$

The lattice parameters can be determined for three different structures bcc, fcc and hcp written as,

$$d = \begin{cases} (3/\pi)^{1/3}\,a & \text{bcc} \\ (3/2\pi)^{1/3}\,a & \text{fcc} \\ (3/\sqrt{3}a^2c/2\pi)^{1/3} & \text{hcp} \end{cases}$$

Equation $n = \dfrac{D^3}{d^3}$ is the expression to calculate cohesive energy for ideal case. It is necessary to introduce a factor k to account for the difference. Therefore,

$$E = k\pi\gamma_o d^2\left(1 - \frac{d}{D}\right)$$

For metals $\dfrac{d}{D}$ is about 10^{-7}. Equation $E = k\pi\gamma_o d^2\left(1 - \dfrac{d}{D}\right)$ can be written as,

$$E_b = k\pi\gamma_o d^2$$

Where E_b is the cohesive energy of the bulk material. When the particle is small, the size of D is in nanometer or smaller d/D is in range of 10^{-2} to 10^{-1}. Rewriting equation $E = k\pi\gamma_o d^2\left(1 - \dfrac{d}{D}\right)$ as,

$$E_p = E_b\left(1 - \frac{d}{D}\right)$$

Where E_p is the cohesive energy of the nanoparticle. Using equation $E_p = E_b\left(1 - \dfrac{d}{D}\right)$ the cohesive energy of nanoparticle can be obtained.

To account for the particle shape difference, let the shape factor be α, which is defined by the equation,

$$\alpha = S'/S$$

where S is the surface area of the spherical nanoparticle and $S = 4\pi R^2$ (R is the radius). S' is the surface area of the nanoparticle in any shape, whose volume is the same as spherical nanoparticle.

From equation $\alpha = S'/S$, the surface area of a nanoparticle in any shape can be written as,

$$S' = \alpha 4\pi R^2$$

Assuming the atoms of nanoparticles are ideal spheres then the contribution to the article surface area of each atoms is πr^2. The number of the surface atoms N is the ratio of the particle surface area to πr^2, which is simplified as $N = 4\alpha (R^2/r^2)$. The volume of the nanoparticle V is the same as the spherical nanoparticle, which equals to $4/3 \pi R^3$. Then the number of the total atoms of the nanoparticle is the ration of the particle volume to the atomic volume $4/3 \pi r^3$ that results to,

$$n = \frac{R^3}{r^3}$$

The cohesive energy of metallic nanoparticle is the sum of the bond energies of all the atoms. Considering equation $E_p = E_b \left(1 - \frac{d}{D}\right)$, the cohesive energy of metallic crystal in any shape (E_p) can be written as,

$$E_p = \frac{1}{2}\left[\frac{1}{4}\beta 4\alpha \frac{R^2}{r^2} + \beta(\frac{R^3}{r^3} - 4\alpha \frac{R^2}{r^2})\right] E_{bond}$$

Where E_{bond} is the bond energy. The value ½ is due to the fact that each bond belongs to two atoms. We can write equation $E_p = \frac{1}{2}\left[\frac{1}{4}\beta 4\alpha \frac{R^2}{r^2} + \beta(\frac{R^3}{r^3} - 4\alpha \frac{R^2}{r^2})\right] E_{bond}$ as,

$$E_p = \tfrac{1}{2} n\beta E_{bond}\left(1 - 6\alpha \frac{r}{D}\right)$$

Where D is the size of the crystal and D = 2R. Rewriting equation $E_p = \tfrac{1}{2} n\beta E_{bond}\left(1 - 6\alpha \frac{r}{D}\right)$ as,

$$E_p = E_o\left(1 - 6\alpha \frac{r}{D}\right)$$

Where $E_o = (½) n\beta E_{bond}$ and E_o is the energy of the solids. The well empirical relation of the melting temperature and the cohesive energy for pure metals are given as,

$$T_{mb} = \frac{0.032}{k_B} E_o$$

Where T_{mb} is the melting temperature of bulk pure metals. Replacing the cohesive energy of solids E_o by general for E_p, then,

$$T_{mb} = \frac{0.032}{k_B} E_o (1 - 6\alpha \frac{r}{D})$$

Equation $T_{mb} = \frac{0.032}{k_B} E_o (1 - 6\alpha \frac{r}{D})$ can be rewritten as:

$$T_{mb} = T_{mb} \left(1 - 6\alpha \frac{r}{D}\right)$$

Equation $T_{mb} = T_{mb} \left(1 - 6\alpha \frac{r}{D}\right)$ is the general equation for the size and shape dependent melting temperature of crystals. The melting temperature of nanoparticles is apparent only when the particle size of smaller than 100 nm. If the particle size is larger than 100 nm, the melting temperature of the particles approximately equals to the corresponding bulk materials, in other words, the melting temperature of nanoparticles is independent of the particle size.

Hall-Petch Equation

Grain-boundary strengthening (or Hall–Petch strengthening) is a method of strengthening materials by changing their average crystallite (grain) size. It is based on the observation that grain boundaries are insurmountable borders for dislocations and that the number of dislocations within a grain has an effect on how stress builds up in the adjacent grain, which will eventually activate dislocation sources and thus enabling deformation in the neighboring grain, too. So, by changing grain size one can influence the number of dislocations piled up at the grain boundary and yield strength. For example, heat treatment after plastic deformation and changing the rate of solidification are ways to alter grain size.

The relation between yield stress and grain size is described mathematically by the Hall–Petch equation:

$$\sigma_y = \sigma_o + \frac{k_y}{\sqrt{d}}$$

Where σ_y is the yield stress, σ_o is a materials constant for the starting stress for dislocation movement (or the resistance of the lattice to dislocation motion), k_y is the strengthening coefficient (a constant specific to each material), and d is the average grain diameter. It is important to note that the H-P relationship is an empirical fit to experimental data, and that the notion that a pileup length of half the grain diameter causes a critical stress for transmission to or generation in an adjacent grain has not been verified by actual observation in the microstructure.

Theoretically, a material could be made infinitely strong if the grains are made infinitely small. This is impossible though, because the lower limit of grain size is a single unit cell of the material. Even then, if the grains of a material are the size of a single unit cell, then the material is in fact amorphous, not crystalline, since there is no long range order, and dislocations cannot be defined in an amorphous material. It has been observed experimentally that the microstructure with the highest yield strength is a grain size of about 10 nm (3.9×10^{-7} in), because grains smaller than this undergo another yielding mechanism, grain boundary sliding Producing engineering materials with this ideal grain size is difficult because only thin films can be reliably produced with grains of this size. In materials having a bi-disperse grain size distribution, for example those exhibiting abnormal grain growth, hardening mechanisms do not strictly follow the Hall-Petch relationship and divergent behavior is observed.

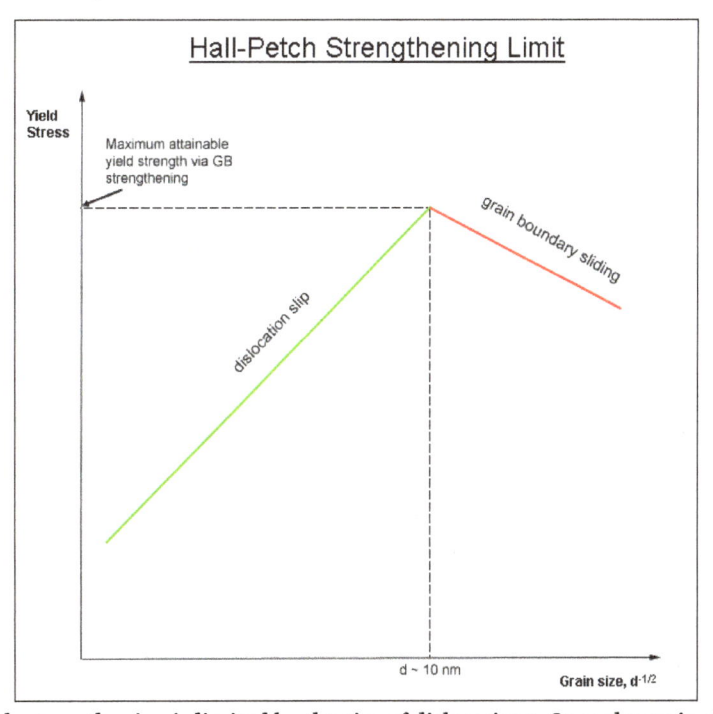

Figure: Hall–Petch strengthening is limited by the size of dislocations. Once the grain size reaches about 10 nanometres (3.9×10^{-7} in), grain boundaries start to slide.

Mechanical Properties

The intrinsic elastic modulus of a nanostructured material is essentially the same as that of the bulk material having micrometer-sized grains until the grain size becomes very small, less than 5 nm. Young's modulus is the factor relating stress and strain. It is the slope of the stress-strain curve in the linear region. The larger the value of Young's modulus, the less elastic the material. Figure below is a plot of the ratio of Young's modulus E in Nano grained iron, to its value in conventional grain-sized iron Eo, as a function of grain size. We see from the figure that below ~20 nm, Young's modulus begins to decrease from its value in conventional grain-sized materials.

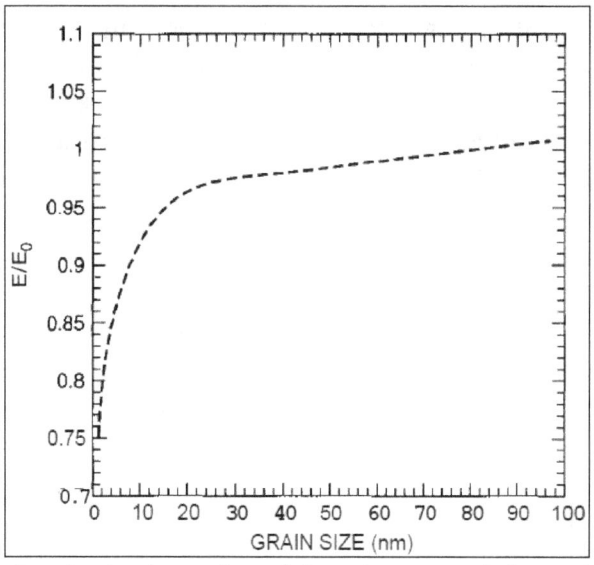

Figure: Plot of ratio of Young's modulus E in Nano grain iron to its value E_o in conventional granular iron as a function of grain size.

Electronic and Optical Properties

The optical and electronic properties of NPs are inter-dependent to greater extent. For instance, noble metals NPs have size dependent optical properties and exhibit a strong UV–visible extinction band that is not present in the spectrum of the bulk metal. This excitation band results when the incident photon frequency is constant with the collective excitation of the conduction electrons and is known as the localized surface plasma resonance (LSPR). LSPR excitation results in the wavelength selection absorption with extremely large molar excitation coefficient resonance Ray light scattering with efficiency equivalent to that of ten fluorophores and enhanced local electromagnetic fields near the surface of NPs that enhanced spectroscopies. It is well established that the peak wavelength of the LSPR spectrum is dependent upon the size, shape and interparticle spacing of the NPs as well as its own dielectric properties and those of its local environment including the substrate, solvents and adsorbates. Gold colloidal NPs are accountable for the rusty colors seen in blemished glass door/windows, while Ag NPs are typically yellow. Actually, the free electrons on the surface in these NPs (d electrons in Ag and gold) are freely transportable through the nanomaterial. The mean free path for Ag and gold is ~50 nm, which is more than the NPs size of these materials. Thus, no scattering is expected from the bulk, upon light interaction, instead they set into a standing resonance conditions, which is responsible for LSPR in these NPs.

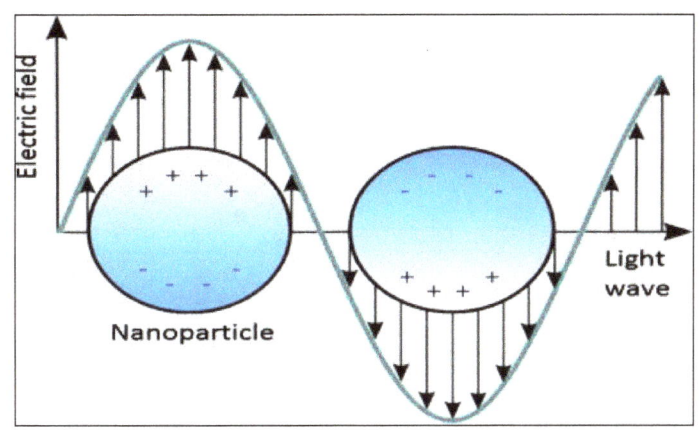

Figure: Graphical illustration exemplifying the localized surface plasmon (LSPR) on nanoparticle outer surface.

Magnetic Properties

Magnetic NPs are of great curiosity for investigators from an eclectic range of disciplines, which include heterogeneous and homogenous catalysis, biomedicine, magnetic fluids, data storage magnetic resonance imaging (MRI), and environmental remediation such as water decontamination. NPs perform best when the size is <critical value i.e. 10–20 nm. At such low scale the magnetic properties of NPs dominated effectively, which make these particle priceless and can be used in different applications. The uneven electronic distribution in NPs leads to magnetic property. These properties are also dependent on the synthetic protocol and various synthetic methods such as solvothermal, co-precipitation, micro-emulsion, thermal decomposition, and flame spray synthesis can be used for their preparation.

Surface Area of Nanoparticles

Nanoparticles have an appreciable fraction of their atoms at the surface, as the data in Tables demonstrate. A number of properties of materials composed of micrometer-sized grains, as well as those composed of nanometre-sized particles, depend strongly on the surface area. For example, the electrical resistivity of a granular material is expected to scale with the total area of the grain boundaries. The chemical activity of a conventional heterogeneous catalyst is proportional to the overall specific surface area per unit volume, so the high areas of nanoparticles provide them with the possibility of functioning as efficient catalysts. It does not follow, however, that catalytic activity will necessarily scale with the surface area in the nanoparticle range of sizes.

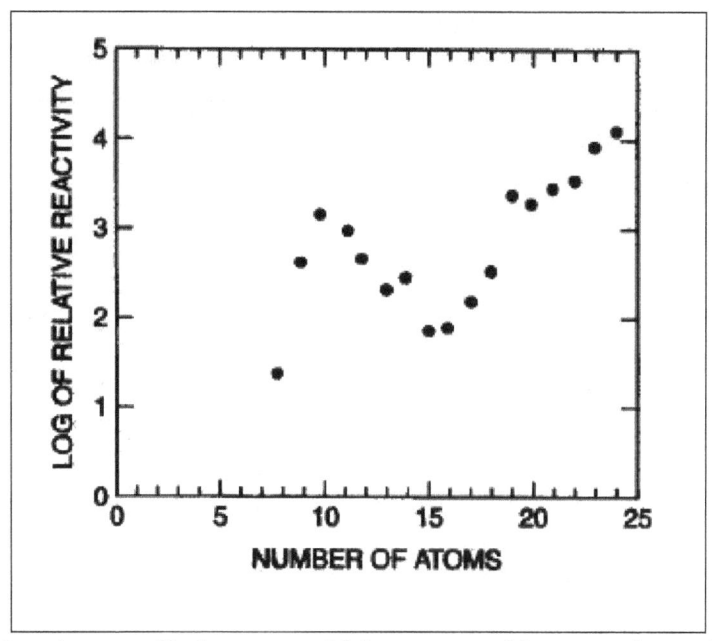

Figure: Reaction rate of hydrogen gas with iron nanoparticles versus the particle size.

Figure above, which is a plot of the reaction rate of H_2 with Fe particles as a function of the particle size, does not show any trend in this direction, and neither does the dissociation rate plotted in figure below for atomic carbon formed on rhodium aggregates deposited on an alumina film. Figure below shows that the activity or turnover frequency (TOF) of the cyclohexene hydrogenation reaction (frequency of converting cyclohexene C_6H_4 to cyclohexane C_6H_6) normalized to the concentration of surface Rh metal atoms decreases with increasing particle size from 1.5 to 3.5 nm, and then begins to level off.

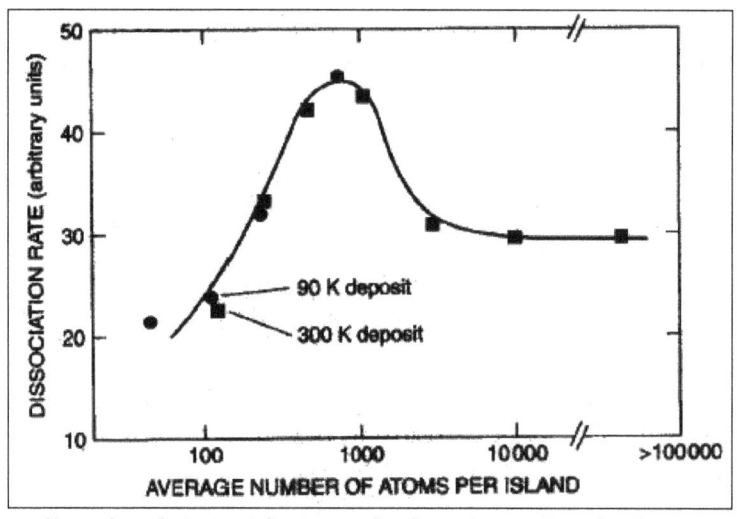

Figure: Effect of catalytic particle size on the dissociation rate of carbon monoxide.

Rhodium aggregates of various sizes, characterized by the number of Rh atoms per aggregate, were deposited on alumina (Al_2O_3) films. The rhodium was given a saturation carbon monoxide (CO) coverage, then the material was heated from 90 to 550 K (circles), or from 300 to 500 K (squares), and the amount of atomic carbon formed on the rhodium provided a measure of the dissociation rate for each aggregate (island) size.

The Rh particle size had been established by the particular alcohol $CnH_{2n+1}OH$ used in the catalyst preparation, where n = 1 for methanol, 2 for ethanol, 3 for 1-propanol, and 4 for 1- butanol. The specific surface area of a catalyst is customarily reported in the units of square meters per gram, denoted by the symbol S, with typical values for commercial catalysts in the range from 100 to 400 m²/g. The general expression for this specific surface area per gram S is,

$$S = \frac{(area)}{\rho(volume)} = \frac{A}{\rho V}$$

Where ρ is the density, which is expressed in the units g/cm³. A sphere of diameter d has the area $A = \pi r^2$ and the volume $V = \pi d^3/6$, to give A/V = 6/d. A cylinder of diameter d and length L has the volume $V = \pi L d^2/4$. The limit L << d corresponds to the shape of a disk with the area $A = \pi d^2/2$, including both sides, to give A/V = 2/d. in like manner, a long cylinder or wire of diameter d and length L >> d has $A = 2\pi rL$, and A/V = 4/V. figures provide the sketches of these figures.

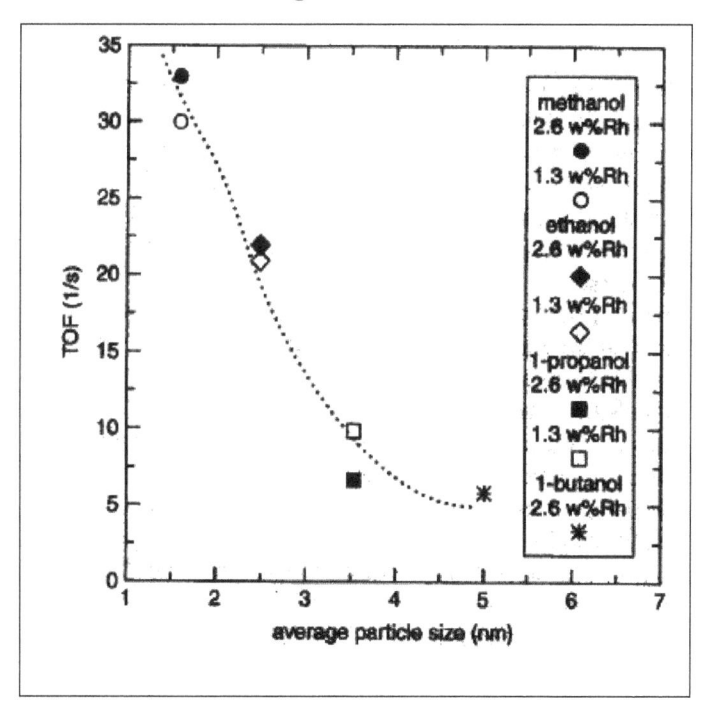

Figure shows the activity of cyclohexane hydrogenation, measured by the turnover frequency (TOF) or rate of conversion of cyclohexane to cyclohexane, plotted as a function of the rhodium (Rh) metal particle size on the surface. The inset gives the alcohols (alkanols) used for the preparation of each particles size.

Using the units square meters per gram, m²/g, for these various geometries we obtain the expressions,

$$S(r) = \frac{6\times 10^3}{\rho d} \text{ sphere of diameter d}$$

$$S(r) = \frac{6\times 10^3}{\rho a} \text{ cube of side a}$$

$$S(r) = \frac{2\times 10^3}{\rho L} \text{ thin disk, } L \ll d$$

$$S(r) = \frac{4\times 10^3}{\rho d} \text{ long cylinder or wire } d \ll L$$

Where the length parameters a, d and L are expressed in nanometres, and the density ρ has the units g/cm³. In equation $S(r) = \frac{2\times 10^3}{\rho L}$ thin disk, $L \ll d$ the area of the side of the disk is neglected, and in equation $S(r) = \frac{4\times 10^3}{\rho d}$ long cylinder or wire $d \ll L$ the areas of the two ends of the wire are disregarded. Similar expressions can be written for distortions of the cube into the quantum-well and quantum-wire configuration of figure.

Table: Densities in g/cm³ of group III-V and II-VI compounds.

	P	As	Sb		S	Se	Te
Al	2.42	3.81	4.22	Zn	4.08	5.42	6.34
Ga	4.13	5.32	5.62	Cd	4.82	5.67	5.86
In	4.79	5.66	5.78	Hg	7.73	8.25	8.17

The densities of type III-V and II-VI semiconductors, from table above are in the range from 2.42 to 8.25 g/cm³, with GaAs having the typical value ρ = 5.32 g/cm³. Using this density we calculated the specific surface area of the nanostructures represented by equations $S(r) = \frac{6\times 10^3}{\rho d}$ sphere of diameter d, $S(r) = \frac{2\times 10^3}{\rho L}$ thin disk, $L \ll d$ and $S(r) = \frac{4\times 10^3}{\rho d}$ long cylinder or wire $d \ll L$, for various values of the size parameters d and L, and the results are represented in the table.

Table: Specific surface areas of GaAs spheres, long cylinders (wires) and thin disks as a function of their size.

Size (nm)	Sphere	Surface area (m²/g) Wire	Disk
4	281	187	94
6	187	125	62
10	112	76	37
20	57	38	19
30	38	26	13
40	29	19	10
60	19	13	6
100	11	8	4
200	9	4	2

The specific surface areas for the smallest structures listed in the table correspond to quantum dots, quantum wires and quantum wells. Their specific surface areas are within the range typical of the commercial catalysts.

The data tabulated above represent minimum specific surface areas in the sense that for a particular mass, or for a particular volume, a spherical shape has the lowest possible area, and for a particular linear mass density, or mass per unit length, a wire of circular cross section has the minimum possible area. It is of interest to study how the specific surface area depends on the shape. Consider a cube of side a with the same volume as a sphere of radius r,

$$\frac{4\pi r^3}{3} = a^3$$

So $a = \left(\frac{4\pi}{3}\right)^{1/3}$

With the aid of equations $S(r) = \frac{6 \times 10^3}{\rho d}$ sphere of diameter d and $S(r) = \frac{6 \times 10^3}{\rho a}$ cube of side a we obtain for this case $S_{cub} = 1.24\ Ss_{ph}$, so a cube has 24% more specific surface area than a sphere with the same volume.

To obtain a more general expression for the shape dependence of the area: volume ratio, we consider a cylinder of diameter D and length L with the same volume as a sphere of radius r, specifically, $4\pi r^3/3 = \pi D^2 L/4$, which gives $r = \frac{1}{2}[3D^2L/2]^{1/3}$. It is easy to show that the specific surface area S(L/D) from equation $S = \dfrac{(\text{area})}{\rho(\text{volume})} = \dfrac{A}{\rho V}$ is given by,

$$S\left(\dfrac{L}{D}\right) = \dfrac{\pi DL + \dfrac{1}{2}\pi D^2}{\rho \pi D^2 \dfrac{L}{4}} = 0.328\, S_{sph}\left[2\left(\dfrac{L}{D}\right)^{1/3}\right] + \left(\dfrac{D}{L}\right)^{2/3}$$

This expression S(L/D), which has a minimum S_{min} = 1.146Ssph for the ratio (LID) = I, is plotted in figure below, normalized relative to Ssph. The normalization factor S_{sph} was chosen because a sphere has the smallest surface area of any object with a particular volume.

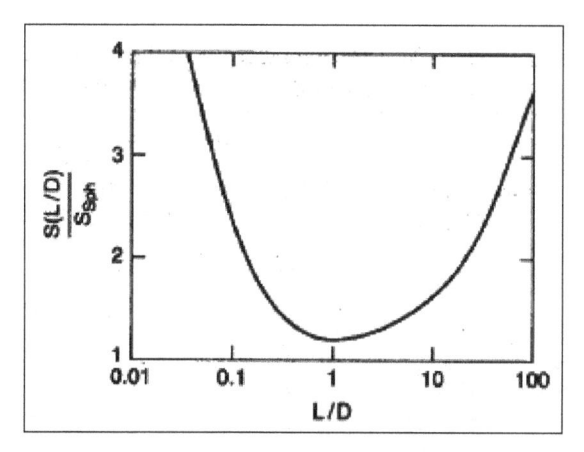

Figure: Dependence of the surface area S(L/D) of cylinder on its length: diameter ratio L/D. the surface area is normalized relative to that of a sphere S_{sph} = 3/ρr with the same volume.

Figure above shows how the surface area increases when a sphere is distorted into the shape of a disk with a particular L/D ratio, without changing in its volume. This figure demonstrates that nanostructures of a particular mass or of a particular volume have much higher surface area S when they are flat or elongated in shape and further distortions rom s regular shape will increase the area even more.

Basic Optical Properties of Nanoparticles

Surface States: Localized Surface Plasmon Resonance of Metal Nanoparticles

The plasmonic effect on metal surface includes two main constituents: surface plasmon polaritons (SPP) and the localized surface plasmon resonance (LSPR). SPP represents the charge density wave propagation on the interface of metal and dielectric, while

LSPR describes the collective oscillation of the conduction electrons in metal nanostructures coupled to electromagnetic field. In the case of SPP, energy is confined within the interface of the metal and the dielectric and propagating in the other two dimensions. It requires necessary phase-matching between the modes from the two sides of the interface. In comparison, the condition to generate LSPR is looser, where plasmon resonances can be excited by direct light illumination. Due to the size restriction of the nanostructure (or nanoparticle), electrons are oscillating collectively against the restoring force of the nuclei, along the direction of the electric field of the electromagnetic wave to which they are coupled, rendering the nanoparticle a dipole-like oscillation source. The analytical description of the LSPR forms the foundation for the optical properties of nanoparticles.

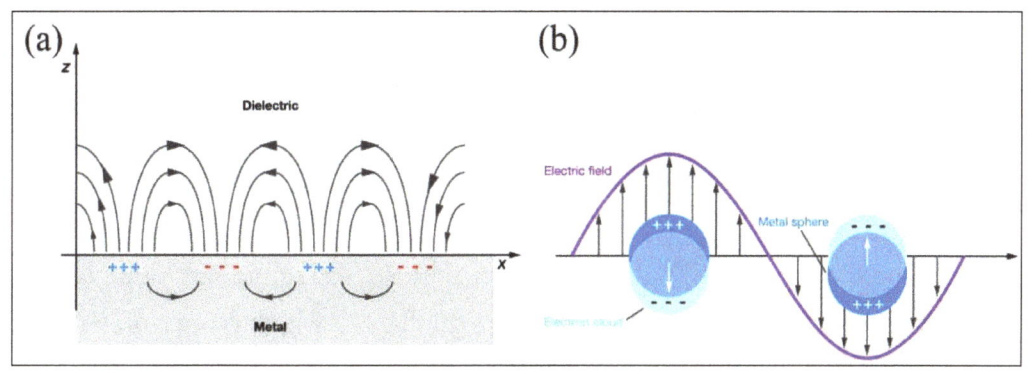

Figure: Schematic diagrams illustrating (a) surface plasmon polaritons (SPP) and (b) localized surface plasmon resonance (LSPR).

To determine the potential inside and around the nanoparticles, the simplest situation, i.e. a homogeneous nano sphere of <100 nm size under the excitation of visible or near-infrared light, is considered. Because the dimension of the nano sphere is much smaller than the wavelength of the light, the electric field throughout the sphere and nearby can be approximated as constant. With such a quasi-static approximation, the general solution to the Laplace equation $\nabla^2 \Phi = 0$ is in the form of,

$$\Phi(r,\theta) = \sum_{l=0}^{\infty} \left[A_l r^l + B_l r^{-(l+1)} \right] P_l(\cos\theta)$$

where θ is as shown in figure above and $P_l(\cos\theta)$ represents the lth order of the Legendre Polynomials. Applying the condition of finite potentials at the origin, one gets,

$$\Phi_{in}(r,\theta) = \sum_{l=0}^{\infty} A_l r^l P_l(\cos\theta);$$

$$\Phi_{out}(r,\theta) = \sum_{l=0}^{\infty} \left[B_l r^l + C_l r^{-1(l+1)} \right] P_l(\cos\theta)$$

where $\Phi_{in}(r,\theta)$ and $\Phi_{out}(r,\theta)$ are potential inside and outside the Nano sphere, respectively. By applying the boundary conditions at the sphere surface and infinitely far away, A_1, B_1 and C_1 can be determined. The potentials can be re-written as:

$$\Phi_{in}(r,\theta) = -\frac{3\varepsilon_b}{\varepsilon(\omega)+2\varepsilon_b} E_o r \cos\theta$$

$$\Phi_{out}(r,\theta) = -E_o r \cos\theta + \frac{\varepsilon(\omega)-\varepsilon_b}{\varepsilon(\omega)+2\varepsilon_b} E_o a^3 r \frac{\cos\theta}{r^2}$$

On the right hand side of $\Phi_{out}(r,\theta) = -E_o r \cos\theta + \frac{\varepsilon(\omega)-\varepsilon_b}{\varepsilon(\omega)+2\varepsilon_b} E_o a^3 r \frac{\cos\theta}{r^2}$, the first term represents the electric field of the excitation, while the second term represents a dipole at the center of the nanoparticle, which is the result of the electron oscillation inside the nanoparticle. Write the second term in the form of $\frac{p \cdot r}{4\pi\varepsilon_o \varepsilon_b r^3}$, where,

$$p = 4\pi\varepsilon_o \varepsilon_b a^3 \frac{\varepsilon(\omega)-\varepsilon_b}{\varepsilon(\omega)+2\varepsilon_b},$$

by the relation $P = \varepsilon_o \varepsilon_b \alpha(\omega) E_o$. It can be seen that for the minimum value of $|\varepsilon(\omega)+2\varepsilon_b|$, there is an enhanced resonance for α. For small or slowly-varying Im[ε(ω)] which is the case for noble metal with visible light, this can be achieved by,

$$\text{Re}[\varepsilon(\omega)] = -2\varepsilon_b.$$

According to the Drude model, $\text{Re}[\varepsilon(\omega)] = -2\varepsilon_b$ is equivalent to,

$$\omega_{sp} = \frac{\omega_p}{(1+2\varepsilon_b)^{1/2}}$$

where ω_p is the bulk plasma frequency of the metal and ω_{sp} is the surface plasmon resonance frequency of the Nano sphere. Equations $\text{Re}[\varepsilon(\omega)] = -2\varepsilon_b$ and $\omega_{sp} = \frac{\omega_p}{(1+2\varepsilon_b)^{1/2}}$ represent the resonance condition (the Fröehlich condition) of the metal sphere, together with other equations above forming the mathematical description of LSPR and its frequency-sensitive feature. As a further step, the electric field inside the Nano sphere and nearby can be derived from $\Phi_{out}(r,\theta) = -E_o r \cos\theta + \frac{\varepsilon(\omega)-\varepsilon_b}{\varepsilon(\omega)+2\varepsilon_b} E_o a^3 r \frac{\cos\theta}{r^2}$ and $\Phi_{in}(r,\theta) = -\frac{3\varepsilon_b}{\varepsilon(\omega)+2\varepsilon_b} E_o r \cos$ as,

$$E_{in} = \frac{3\varepsilon_b}{(\varepsilon+2\varepsilon_b)} E_o;$$

$$E_{out} = E_o + \frac{3n(n.p)-p}{4\pi\varepsilon_o\varepsilon_b}\frac{1}{r^3}$$

It can be seen that the filed both inside the Nano sphere and nearby is enhanced. This property provides the prerequisite for most of the LSPR based applications and devices.

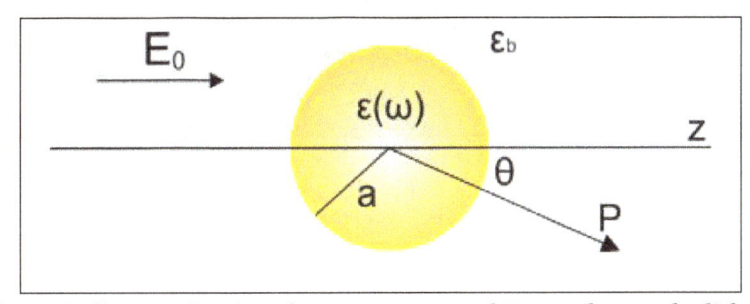

Figure: Schematic diagram showing a homogeneous metal Nano sphere under light excitation.

Nanoparticle-induced Heating

The dipole-like electrons oscillation in LSPR is coupled to electromagnetic field of the excitation source, making nanoparticles efficient absorbers to light with specific frequency. The energy absorbed by the nanoparticles from the excitation can be further released or transferred afterwards. In some cases, the energy is radiatively redistributed to the far field. In some other cases, it is also possible for the energy to be released locally as heat dissipation, due to the damping of the oscillation. Such nanoparticle-induced heat gives rise to temperature increase both within the nanoparticles and nearby. It is termed as the photothermal effect of metal nanoparticles.

Evidence for the photothermal effect is shown in Figure below, which is comprised of three images by different imaging techniques of the same region of a sample on which three different types of nanosperes (300-nm-diameter latex, 80-nm-diameter gold, and 10-nmdiameter gold) are located. In the differential interference contrast image which is not thermal sensitive, the 300-nm spheres are clearly seen, while 80-nm ones are resolved marginally. The 10-nm spheres are not resolved due to the diffraction limit. In comparison, with a thermal sensitive imaging technique, the latex spheres become invisible at all, while the gold spheres are detectable. With a high excitation power, even the 10-nm ones can be resolved. It is enlightened from the above results that the photothermal effects of metal nanoparticles can be employed to image objects that is under the diffraction limit of the illuminating light. It is based on the fact that the optical absorption of metal nanoparticles decreases slower as the particles size shrinks down (to the third power), than that of the scattering does (to the sixth power), resulting in the domination of the photothermal effect over the scattering among small particles. Furthermore, metal nanoparticles are not photo bleached and show no optical saturation with moderate excitation power, which make them very attractive for the imaging applications.

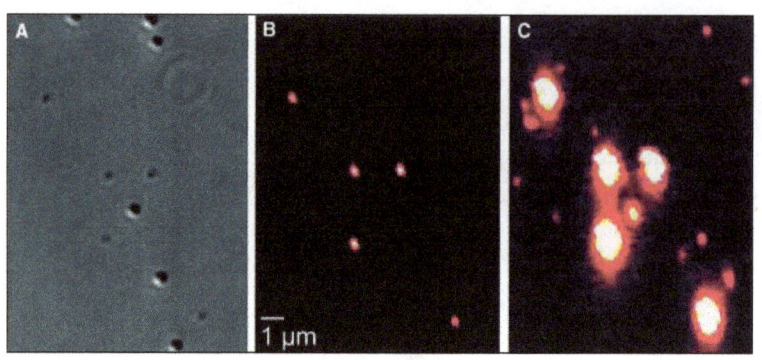

Figure: Images, by non-thermal-sensitive (a) and thermal-sensitive (b) and (c) imaging techniques, of the same area of a sample containing three different types of nanospheres (300-nm-diameter latex, 80-nmdiameter gold, and 10-nm-diameter gold).

Apart from imaging techniques, the photothermal effect also helps metal nanoparticles to find applications in medical therapy again cancers. The local temperature increase caused by nanoparticle-induced heat has proved to be effective for selective destruction of cancer cells. Cancer cells are more sensitive to temperature alteration due to the higher metabolic rates they have, compared with normal cells. Therefore, it is possible to control the temperature increase quantitatively so that it is enough to kill the cancer cells without damaging other normal cells. The precise control of the temperature is one key issue for this technique. It also forms one of the subjects for the original work of the thesis. For in vivo therapies, gold nanospheres are not competent because of absorption outside the near infrared window in biological tissues (650 nm to 900 nm). Instead, gold nanoshells coated on dielectric cores and nanorods with high aspect ratio are introduced, utilizing the appropriate absorption band of these structures.

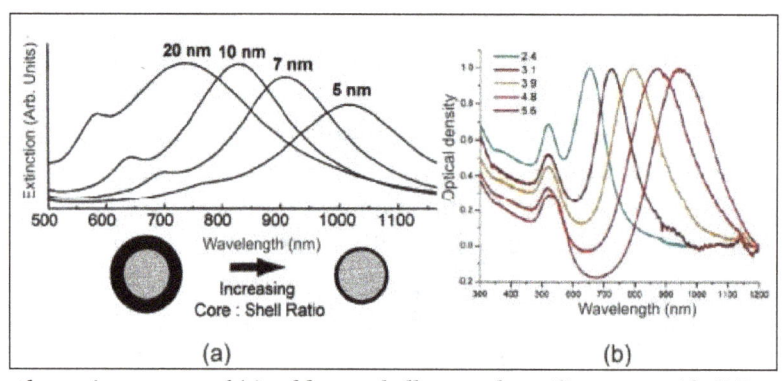

Figure: Absorption spectra of (a) gold nanoshells coated on silica cores with different shell thickness, and (b) gold nanorods with different aspect ratios.

There is a tendency in recent years to combine the high competence of photothermal effects of metal nanoparticles in imaging techniques and medical therapy together. On one hand, nanoparticles are used as imaging agents for different diagnostic approaches like MRI, CT, and Optical imaging. On the other hand, they are functioning as therapeutic tools at the same time.

Interaction between Metal Nanoparticles and Fluorophores

Metal nanoparticles are able to considerably magnify the electric field both within the particles and confined near the surface, due to LSPR. Therefore, they are introduced to the vicinity of fluorophores (e.g. dye molecules, quantum dots, etc.) to modify optical properties (most importantly the fluorescence rate or fluorescence intensity) of the fluorophores; a technique named as radiative decay engineering. The most commonly used nanoparticles in this technique is made of noble metals such as gold and silver, because of the appropriate LSPR band in the visible and near-infrared range by these metals. The fluorescence intensity of the fluorophores can possibly be either enhanced or quenched, determined by a variety of factors like the fluorophore-nanoparticle distance, the size and shape of the nanoparticles, the dipole momentum orientation of the fluorophores and the nanoparticles, etc. The metal nanoparticles exert the optical properties modification onto the fluorophores via three main interactions: the increase in the excitation rate, the modification in the radiative decay rate, together with the non-radiative energy transfer. The interactions can be expressed with the simplest relation as,

$$\gamma_{em} = \gamma_{exc} \frac{\gamma_r}{\gamma_r + \gamma_{nr}}$$

where γ_{em}, γ_{exc}, γ_r and γ_{nr} represents the fluorescence rate, the excitation rate, the radiative decay rate, and the non-radiative rate, respectively. The term $q_a = \gamma_{exc} \frac{\gamma_r}{\gamma_r + \gamma_{nr}}$ is also called the quantum yield of the fluorophores. The first two interactions are likely to enhance the fluorescence intensity, while the last one leads to fluorescence quenching. A fourth type of interaction, namely the emission redirection, is also discussed elsewhere.

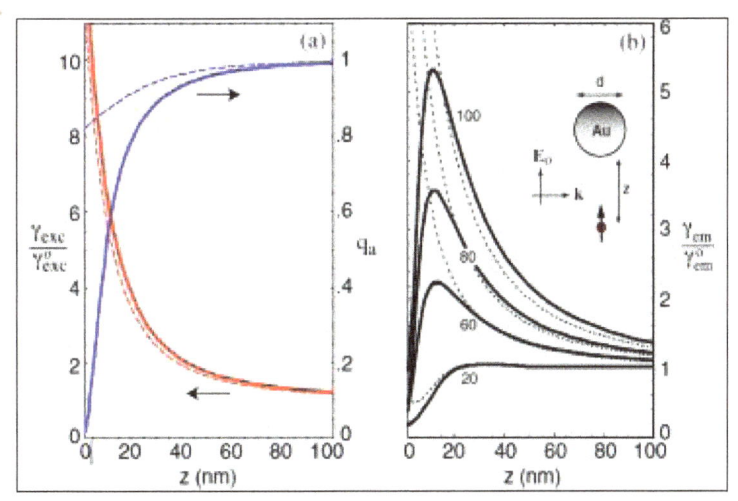

Figure: Calculated (a) excitation rate enhancement, quantum yield, and (b) fluorescence rate modification of a fluorescent molecule by a single spherical gold nanoparticle of 80 nm diameter as a function of the distance between the molecule and the nanoparticle surface.

The solid curves and the dashed curves are corresponding to different calculation methods of the multiple F be discussed later multipoles (MMP) method and the dipole approximation, respectively.

1. To increase the excitation rate γ_{exc}. The excitation rate of a fluorescent molecule is proportional to the local excitation field ($\gamma_{exc} \propto p\,E\cdot$, with p representing the transition dipole momentum of the molecule). The presence of a metal nanoparticle gives rise to field enhancement near the surface. Thus it is able to increase the excitation rate of a fluorescent molecule in the vicinity of the nanoparticle. The red curves show the calculated excitation rate of a molecule which is located with distance z to the surface of a single spherical gold nanoparticle of diameter d = 80 nm, with respect to that of the same molecule in free space (without the nanoparticle). Because the size of the molecule is much smaller than d, it is treated as a point light source. It is seen that the enhancement of the excitation rate is over one order of magnitude in the very near region to the nanoparticle surface, due to the profoundly magnified electric field in this region by LSPR.

The small discrepancy between the two red curves comes from the different methods used for calculation. The dashed curve is obtained with the dipole approximation of the gold nanoparticle, while the solid curve is the result of the multiple multipole (MMP) method with which higher orders are taken into account as well. It is obvious that the solid curve is more accurate than the dashed one. However, the difference between the calculations of the excitation rate enhancement is negligible, due to the fact that the dipole approximation holds very well under the circumstance here. When two conditions are met: (1) d << λ, and (2) the electric field of the excitation source is homogeneous across the nanoparticle. Condition (1) is obviously met here, as λ is the wavelength of the excitation in this case. Condition (2) is also true, since the distance from the excitation source to the nanoparticle is much larger than the dimension of the nanoparticle and the excitation light can be treated as plane wave. Therefore, the dipole approximation is valid for the calculation of γ_{exc}.

2. To modify the radiative decay rate γ_r. According to the Purcell effect, the rate for spontaneous emission (γ_r) is not an intrinsic property of a fluorophore. It is also affected by the electromagnetic field of the ambient environment. In the current scenario, the concentrated light field near the surface of a nanoparticle by LSPR is able to alter the radiative decay rate of a fluorophore within this region. It is fundamentally a result of the high local density of optical states near the nanoparticle surface, which increases the sum of the radiative and non-radiative decay rate of the fluorophore. Notably, the increase in the total decay rate does not necessarily mean an increase in any of the two parts (γ_r and γ_{nr}).

In some cases, an increase in one part combined with a decrease in the other part is also possible. How the modification as a sum is allocated into the two parts is determined by the distance between the fluorophore and the nanoparticle. When the distance is small (less than 10 nm), the increase in the non-radiative decay rate is dominating, by a non-radiative transfer of energy from the fluorophore to the nanoparticle, which will be discussed later. It is worth noting that the dipole approximation which is suitable for the excitation rate enhancement calculation is no longer valid here for calculating the modification of the decay rates. This is simply because the distance from the nanoparticle to the excitation source (the fluorophore in this case) is comparable to the dimension of the nanoparticles so that the second condition for the dipole approximation (homogeneous field across the nanoparticle) is violated.

Energy Transfer

The increase in the non-radiative decay rate of a fluorescent molecule due to the concentrated light field by a metal nanoparticle nearby is realized as an energy transfer from the excited molecule to the nanoparticle. The energy transfer is non-radiative and competes with other radiative transitions. Therefore it is likely to suppress the quantum yield of the fluorescent molecule. On the other hand, the rate of the energy transfer is molecule-nanoparticle separation dependent. It is found that this is a typical Förster resonance energy transfer (FRET) process. The energy transfer rate is inversely proportional to the sixth power of the distance between the molecule and the nanoparticle, causing a prominent domination of the transfer within the near region to the nanoparticle surface ($z < 10$ nm). The quantum yield modification of the fluorophore by the gold nanosphere with the two blue curves.

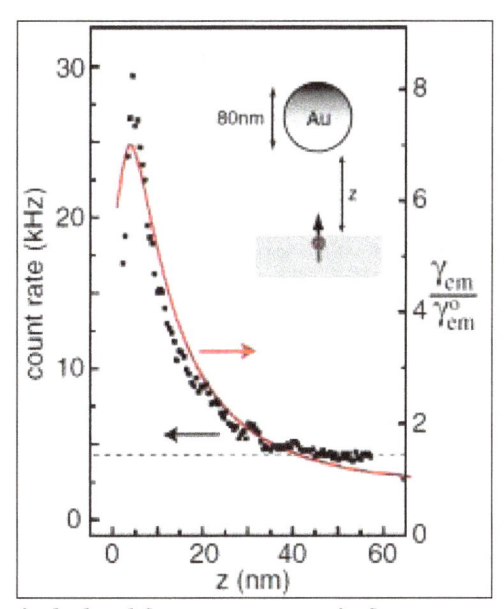

Figure: Comparison of calculated fluorescence rate of a fluorescent molecule as a function of the distance to the surface of an 80 nm-diameter gold nanoparticle (the red curve) with experimental data (the black squares).

It is clear that calculation based on the dipole approximation (the dashed blue curve) fails to describe dominating non-radiative decay at short distances due to strong energy transfer, while the MMP method gives more accurate result (the solid blue curve). The energy transfer near the very surface of the nanosphere can be so pronounced that the more than ten times enhancement in the excitation rate is cancelled out, resulting in quenching of the fluorescence intensity. In the same figure, it is once again demonstrated that the dipole approximation is not accurate enough by the fact that fluorescence quenching is not predicted (the dashed curves). Of practical interests, the energy transfer is possible to be suppressed, where it is not favored, by coating insulating materials like silica, since the transfer is realized by the form of electronic energy transfer. The coating material functions both as an insulating layer to deactivate the transfer process and as a spacer to keep the fluorophore in a 'safe' distance away from the nanoparticle. However, the energy transfer can be favorable under some circumstances, e.g. for imaging techniques with the metal nanoparticles as contrast agents, for Raman spectroscopy, or for photovoltaic applications.

Figure shows the comparison of calculated fluorescence rate of the fluorescent molecule to that obtained by experimental measurement. It is seen that the calculation result agrees well with the experimental data, thus demonstrating the high accuracy of the model comprised of the three interactions. In practice, there still exists another type of interaction between fluorophores and metal nanoparticles, e.g. the destruction of the fluorophores by the heat generated by nanoparticles with photothermal effect. The principle is similar to that of nanoparticle-involved photothermal therapy. The destruction is exerted onto the chemical structure of the fluorophores, and is an irreversible process in most of the cases. Therefore this type of interaction is not counted for radiative decay engineering. Apparently the photothermal effects are not welcomed here for RDE, in contrast to the situation for photothermal therapeutic applications. To restrain the effect, silica coating can also be applied, due to the high thermal resistance of the material.

Photoluminescence

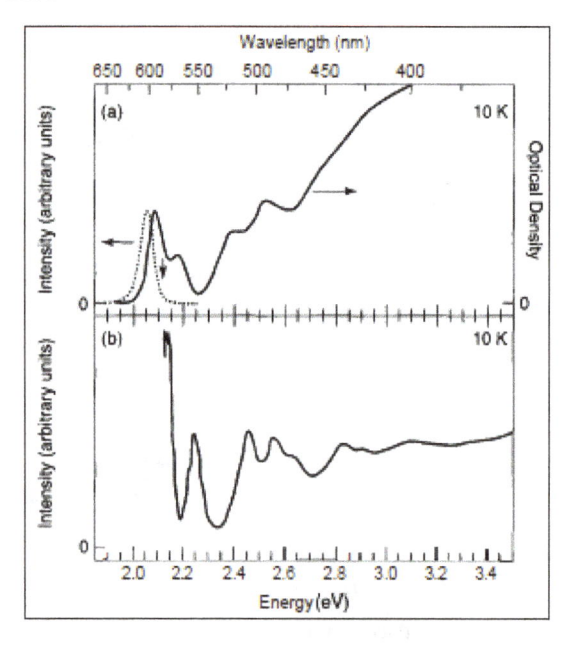

Figure shows spectra taken at 10 K for 5.6 nm diameter CdSe quantum dots: (a) absorption spectrum (solid line) and photoluminescence spectrum (dashed line) obtained with excitation at 2.655 eV (467 nm); (b) photoluminescence spectrum obtained with the emission position marked by the downward arrow on the upper plot.

The technique of photoluminescence excitation (PLE) has become a standard one for obtaining information on the nature of nanostructures such as quantum dots. In bulk materials the luminescence spectrum often resembles a standard direct absorption spectrum, so there is little advantage to studying the details of both. High photon excitation energies above the band gap can be the most effective for luminescence studies of bulk materials, but it has been found that for the case of nanoparticles the efficiency of luminescence decreases at high incoming photon energies. Non-radiative relaxation pathways can short-circuit the luminescence at these high energies, and it is of interest to investigate the nature of these pathways.

The photoluminescence excitation technique involves scanning the frequency of the excitation signal, and recording the emission within a very narrow spectral range. Figure illustrates the technique for the case of -5.6-nm CdSe quantum dot nanoparticles. The solid line in Figure above plots the absorption spectrum in the range from 2.0 to 3.1 eV and the superimposed dashed line shows the photoluminescence response that appears near 2.05 eV. The sample was then irradiated with a range of photon energies of 2.13-3.5eY and the luminescence spectrum emitted at the photon energy of 2.13eV is shown plotted in Figure b as a function of the excitation energy. The downward-pointing arrow on Figure above indicates the position of the detected luminescence. It is clear from a comparison of the absorption and luminescence spectra of this figure that the photoluminescence (b) is much better resolved.

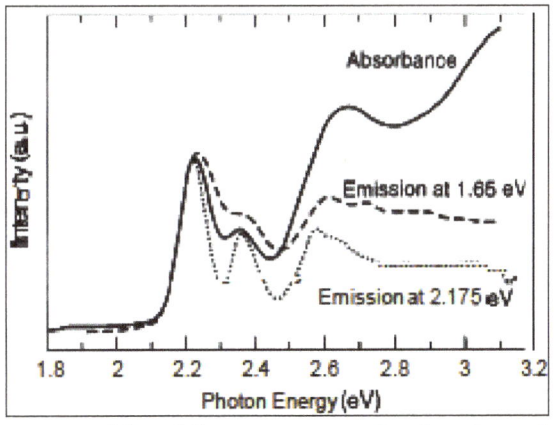

Figure: Spectra for CdSe nanoparticles of diameter 3.2 nm, showing absorption spectrum (solid line), excitation spectrum for emission at the 2.175 eV band edge fluorescence maximum (dark dashed line), and excitation spectrum for emission at the 1.65 eV deep-trap level (light dashed line).

The excitation spectra of nanoparticles of CdSe with a diameter of 3.2 nm exhibit the expected band edge emission at 2.176 eV at the temperature 77 K, and they also exhibit an emission signal at 1.65 eV arising from the presence of deep traps. Figure above compares the PLE spectra for the band-edge and deep-trap emissions with the corresponding absorption, and we see that the band-edge emission is much better resolved. This is because each particle size emits light at a characteristic frequency so the PLE spectrum reflects the emission from only a small fraction of the overall particle size distribution. Shallow traps that can be responsible for band-edge emission have the same particle size dependence in their spectral response. This considerably reduces the inhomogeneous broadening, and the result is a narrowed, nearly homogeneous spectrum. The emission originating from the deep traps does not exhibit this same narrowing, which explains the low resolution of the 1.65 eV emission spectrum.

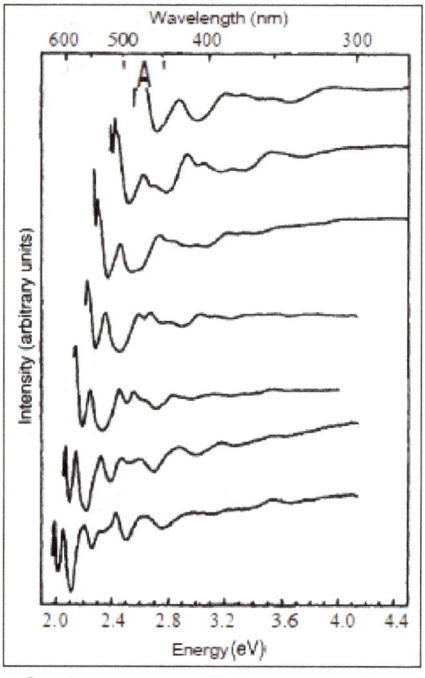

Figure: Normalized photoluminescence excitation spectra for seven CdSe quantum dots ranging in size from -1.5 nm (top spectrum) to -4.3 nm (bottom spectrum).

We mentioned above that there is a shift, that is, a shift of spectral line positions to higher energies as the size of a nanoparticle decreases. This is dramatically illustrated by the photoluminescence emission spectra presented in Figure arising from seven quantum dot samples ranging in size from - 1.5 nm for the top spectrum to -4.3 nm for the bottom spectrum. We see that the band edge gradually shifts to higher energies, and the distances between the individual lines also gradually increase with the decrease in particle size. Another way to vary spectral parameters is to excite the sample with a series of photon energies and record the fluorescence spectrum over a range of energies, and this produces the series of spectra illustrated in Figure below. On this figure the peak of the fluorescence spectrum shifts to higher energies as the excitation photon energy increases. We also notice from the absorption spectrum, presented at the bottom of the figure for comparison purposes that for all photon excitation energies the fluorescence maximum is at lower energies than the direct absorption maximum.

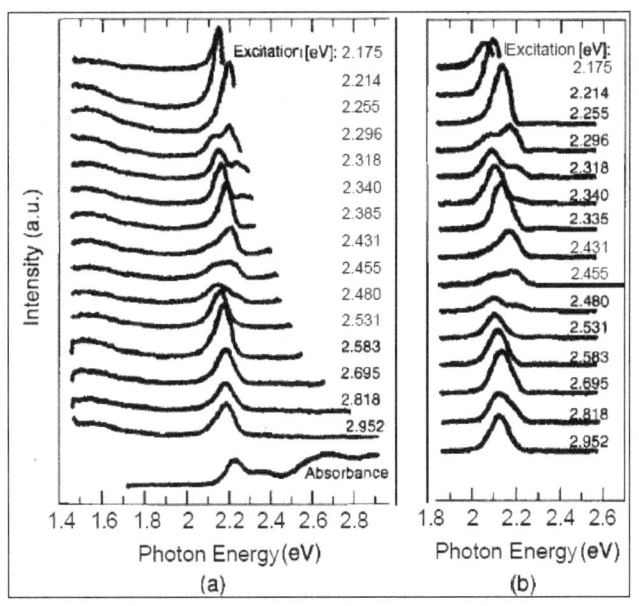

Figure: Fluorescence spectra for 3.2 nm diameter CdSe nanocrystals for various indicated excitation nergies at 77K: (a) experimental and (b) simulated spectra. For comparison purposes, the experimental absorption spectrum is shown at the bottom left.

Thermoluminescence

Another spectral technique that can provide information on surface states, detrapping, and other processes involved in light emission from nanoparticles is thermo-luminescence, the emission of light brought about by heating. Sometimes electron-hole pairs produced by irradiating a sample do not recombine rapidly, but become trapped in separate metastable states with prolonged lifetimes. The presence of traps is especially pronounced in small nanoparticles where a large percentage of the atoms are at the surface, many with unsatisfied chemical bonds and unpaired electrons.

Heating the sample excites lattice vibrations that can transfer kinetic energy to electrons and holes held at traps, and thereby release them, with the accompaniment of emitted optical photons that constitute the thermal luminescence. To measure thermo-luminescence, the energy needed to bring about the release of electrons and holes from traps is provided by gradually heating the sample and recording the light emission as a function of temperature for CdS residing in the cages of the material zeolite-Y. The energy corresponding to the maximum emission, called the glow peak, is the energy needed to bring about the detrapping, and it may be considered as a measure of the depth of the trap. This energy, however, is generally insufficient to excite electrons from their ground states to excited states. For example, at room temperature (300 K) the thermal energy k_BT= 25.85 meV is far less than typical gap energies E_g, although it is comparable to the ionization energies of many donors and acceptors in semiconductors. It is quite common for trap depths to be in the range of thermal energies.

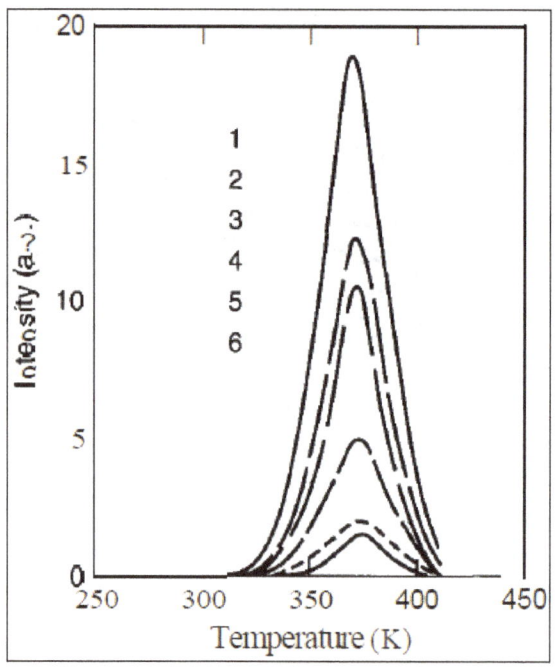

Figure: Glow curves of CdSe clusters in zeolite-γ for CdS loadings of 1, 3, 5, and 20% (curves 1-4 respectively). Curves 5 is for bulk CdS and curve 6 is for a mechanical mixture of CdS with zeolite-γ powder.

Synthesis of Nanoparticles

Various methods can be employed for the synthesis of NPs, but these methods are broadly divided into two main classes i.e. (1) Bottom-up approach and (2) Top-down approach as shown in figure below. These approaches further divide into various sub-classes based on the operation, reaction condition and adopted protocols.

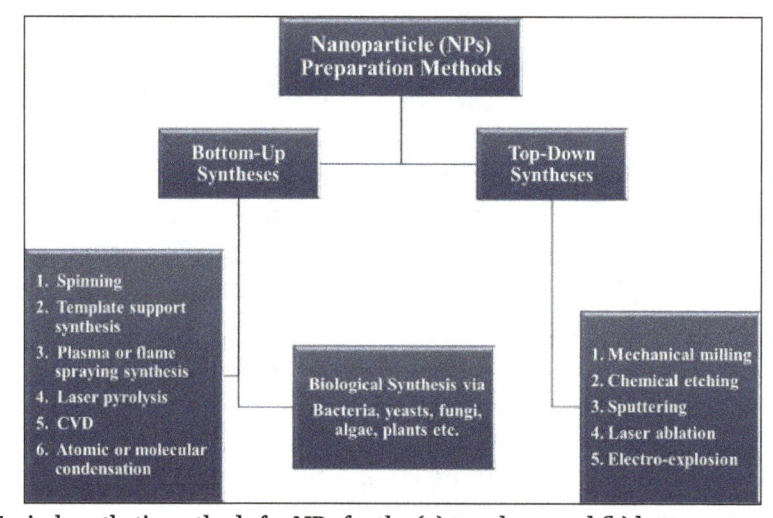

Figure: Typical synthetic methods for NPs for the (a) top-down and (b) bottom-up approaches.

Top-down Synthesis

In this method, destructive approach is employed. Starting from larger molecule, which decomposed into smaller units and then these units are converted into suitable NPs. Examples of this method are grinding/milling, CVD, physical vapor deposition (PVD) and other decomposition techniques. This approach is used to synthesize coconut shell (CS) NPs. The milling method was employed for this purpose and the raw CS powders were finely milled for different interval of times, with the help of ceramic balls and a well-known planetary mill. They showed the effect of milling time on the overall size of the NPs through different characterization techniques. It was determined that with the time increases the NPs crystallite size decreases, as calculated by Scherer equation. They also realized that with each hour increment the brownish color faded away due to size decrease of the NPs. The SEM results were also in an agreement with the X-ray pattern, which also indicated the particle size decreases with time.

the spherical magnetite NPs synthesis from natural iron oxide (Fe_2O_3) ore by top-down destructive approach with a particle size varies from ~20 to ~50 nm in the presence of organic oleic acid. A simple top-down route was employed to synthesize colloidal carbon spherical particles with control size. The synthesis technique was based on the continuous chemical adsorption of polyoxometalates (POM) on the carbon interfacial surface. Adsorption made the carbon black aggregates into relatively smaller spherical particles, with high dispersion capacity and narrow size distribution as shown in Figure below. It also revealed from the micrographs, that the size of the carbon particles becomes smaller with sonication time. A series of transition-metal dichalcogenide nanodots (TMD-NDs) were synthesized by combination of grinding and sonication top-down techniques from their bulk crystals. It was revealed that almost all the TMD-NDs with sizes <10 nm show an excellent dispersion due to narrow size distribution. Lately, highly photoactive active Co_3O_4 NPs were prepared via top-down laser fragmentation, which is a top-down process.

Bottom-up Synthesis

This approach is employed in reverse as NPs are formed from relatively simpler substances, therefore this approach is also called building up approach. Examples of this case are sedimentation and reduction techniques. It includes sol gel, green synthesis, spinning, and biochemical synthesis.

TiO_2 anatase NPs with graphene domains through this technique was synthesized. Alizarin and titanium isopropoxide precursors to synthesize the photoactive composite for photocatalytic degradation of methylene blue were used. Alizarin was selected as it offers strong binding capacity with TiO_2 through their axial hydroxyl terminal groups. The anatase form was confirmed by XRD pattern. The SEM images taken for different samples with reaction scheme are provided in figure below. SEM indicates that with temperature elevation, the size of NPs also increases.

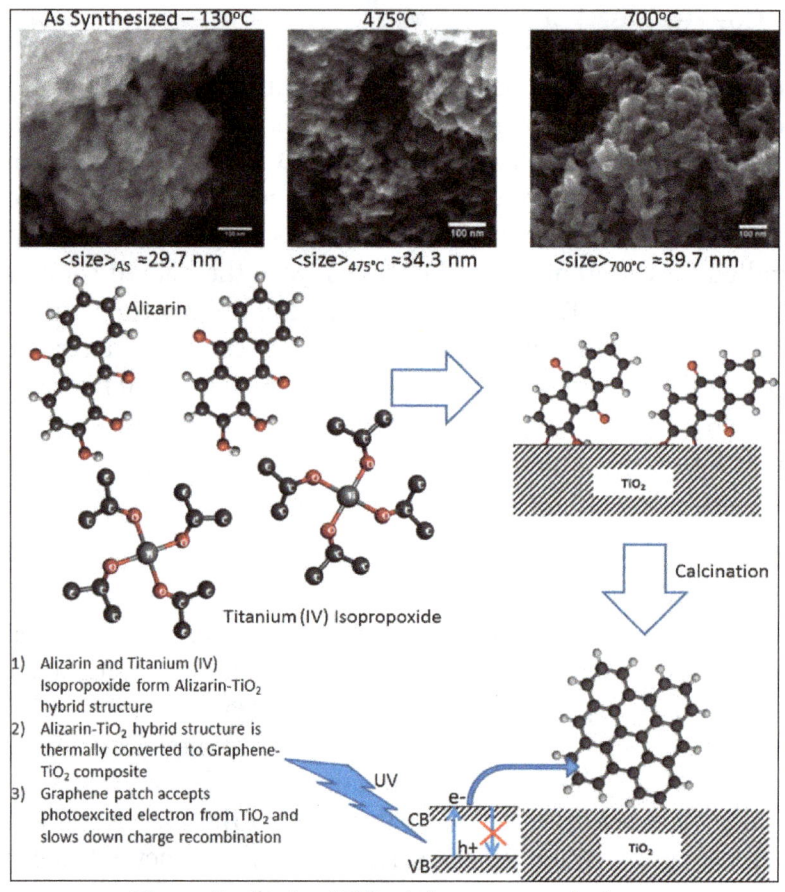

Figure: Synthesis of TiO$_2$ via bottom-up technique.

Well-uniform spherical shaped Au nanospheres with monocrystalline have been synthesized via laser irradiation top-down technique.

Chill Block Melting Method

Nanostructured materials can be made by rapid solidification. One method illustrated in figure is called "chill block melt spinning." RF (radiofrequency) heating coils are used to melt a metal, which is then forced through a nozzle to form a liquid stream. This stream is continuously sprayed over the surface of a rotating metal drum under an inert-gas atmosphere. The process produces strips or ribbons ranging in thickness from 10 to 100 pm. The parameters that control the nanostructure of the material are nozzle size, nozzle-to drum distance, melt ejection pressure, and speed of rotation of the metal drum. The need for light weight, high strength materials has led to the development of 85-94% aluminum alloys with other metals such as Y, Ni, and Fe made by this method. A melt spun alloy of Al-Y-Ni-Fe consisting of 10-30-nm Al particles embedded in an amorphous matrix can have a tensile strength in excess of 1.2 GPa. The high value is attributed, to the presence of defect free aluminum nanoparticles.

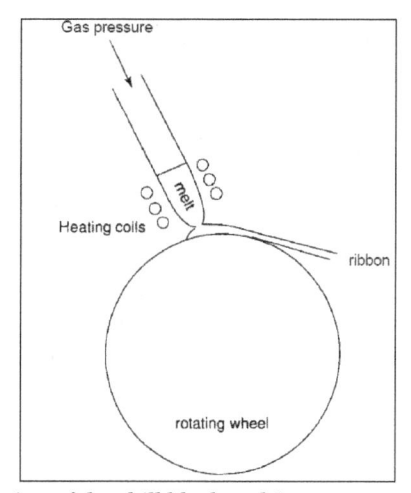

Figure: Illustration of the chill block melting apparatus for producing nanostructured materials by rapid solidification on a rotating wheel.

Gas Atomization

In another method of making nanostructured materials, called gas atomization, a high velocity inert-gas beam impacts a molten metal. The apparatus is illustrated in Figure below. A fine dispersion of metal droplets is formed when the metal is impacted by the gas, which transfers kinetic energy to the molten metal. This method can be used to produce large quantities of nanostructured powders, which are then subjected to hot consolidation to form bulk samples. Nanostructured materials can be made by electrode position. For example, a sheet of nanostructured Cu can be fabricated by putting two electrodes in an electrolyte of $CuSO_4$ and applying a voltage between the two electrodes. A layer of nanostructured Cu will be deposited on the negative titanium electrode. A sheet of Cu_2 mm thick can be made by this process, having an average grain size of 27 nm, and enhanced yield strength of 119 MPa.

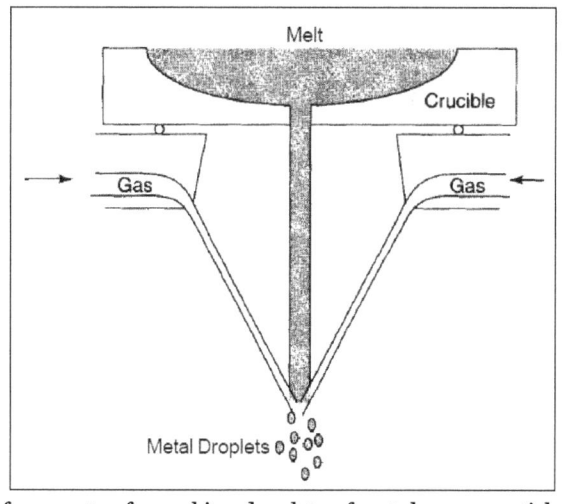

Figure: Illustration of apparatus for making droplets of metals nanoparticles by gas atomization.

Comparison between Top-down and Bottom-up Synthetic Techniques

Top–down method	Merits	Demerits	General remarks
Optical lithography	Long-standing, established micro/nanofabrication tool especially for chip production, sufficient level of resolution at high throughputs.	Tradeoff between resist process sensitivity and resolution, involves state-of-the-art expensive clean room based complex operations.	The 193 nm lithography infrastructure already reached a certain level of maturity and sophistication, and the approach could be extended to extreme ultraviolet (EUV) sources to shrink the dimension. Also, future developments need to address the growing cost of a mask set.
E-beam lithography	Popular in research environments, an extremely accurate method and effective nanofabrication tool for <20 nm nanostructure fabrication with desired shape.	Expensive, low throughput and a slow process (serial writing process), difficult for <5 nm nanofabrication.	E-beam lithography beats the diffraction limit of light, capable of making periodic nanostructure features. In the future, multiple electron beam approaches to lithography would be required to increase the throughput and degree of parallelism.
Soft and nanoimprint lithography	Pattern transfer based simple, effective nanofabrication tool for fabricating ultra-small features (<10 nm).	Difficult for large-scale production of densely packed nanostructures, also dependent on other lithography techniques to generate the template, and usually not cost-effective.	Self-assembled nanostructures could be a viable solution to the problem of complex and costly template generation, and for templates of periodic patterns of <10 nm.
Block co-polymer lithography	A high-throughput, low-cost method, suitable for large-scale densely packed nanostructures, diverse shapes of nanostructures, including spheres, cylinders, lamellae possible to fabricate including parallel assembly.	Difficult to make self-assembled nanopatterns with variable periodicity required for many functional applications, usually high defect densities in block copolymer self-assembled patterns.	Use of triblock copolymers is promising to generate more exotic nanopattern geometries. Also, functionalization of parts of the block copolymer could be done to achieve hierarchy of nanopatterning in a single step nanofabrication process.
Scanning probe lithography	High resolution chemical, molecular and mechanical nanopatterning capabilities, accurately controlled nanopatterns in resists for transfer to silicon, ability to manipulate big molecules and individual atoms.	Limited for high throughput applications and manufacturing, an expensive process, particularly in the case of ultra-high-vacuum based scanning probe lithography.	Scanning probe lithography can be leveraged for advanced bionanofabrication that involves fabrication of highly periodic biomolecular nanostructures.

Bottom–up method	Merits	Demerits	General remarks
Atomic layer deposition	Allows digital thickness control to the atomic level precision by depositing one atomic layer at a time, pin-hole free nanostructured films over large areas, good reproducibility and adhesion due to the formation of chemical bonds at the first atomic layer.	Usually a slow process, also an expensive method due to the involvement of vacuum components, difficult to deposit certain metals, multicomponent oxides, certain technologically important semiconductors (Si, Ge, etc.) in a cost-effective way.	Although a slow process, it is not detrimental for the fabrication of future generation ultra-thin ICs. The stringent requirements for the metal barriers (pure; dense; conductive; conformal; thin) that are employed in modern Cu-based chips can be fulfilled by atomic layer deposition.
Sol gel nanofabrication	A low-cost chemical synthesis process based method, fabrication of a wide variety of nanomaterials including multicomponent materials (glass, ceramic, film, fiber, composite materials).	Not easily scalable, usually difficult to control synthesis and the subsequent drying steps.	A versatile nanofabrication method that can be made scalable with further advances in the synthesis steps.
Molecular self-assembly	Allows self-assembly of deep molecular nanopatterns of width less than 20 nm and with the large pattern stretches, generates atomically precise nanosystems.	Difficult to design and fabricate nanosystems unlike mechanically directed assembly.	Molecular self-assembly of multiple materials may be an useful approach in developing multifunctional nanosystems and devices.
Physical and chemical vapor-phase deposition	Versatile nanofabrication tools for fabrication of nanomaterials including complex multicomponent nanosystems (e.g. nanocomposites), controlled simultaneous deposition of several materials including metal, ceramics, semiconductors, insulators and polymers, high purity nanofilms, a scalable process, possibility to deposit porous nanofilms.	Not cost-effective because of the expensive vacuum components, high-temperature process and toxic and corrosive gases particularly in the case of chemical vapor deposition.	It provides unique opportunity of nanofabrication of highly complex nanostructures made of distinctly different materials with different properties that are not possible to accomplish using most of the other nanofabrication techniques. New advances in chemical vapor deposition such as 'initiated chemical vapor deposition' (i-CVD) provide unprecedented opportunities of depositing polymers without reduction in the molecular weights.
DNA-scaffolding	Allows high-precision assembling of nanoscale components into programmable arrangements with much smaller dimensions (less than 10 nm in half-pitch).	Many issues need to explore, such as novel unit and integration processes, compatibility with CMOS fabrication, line edge roughness, throughput and cost.	Very early stage. Ultimate success depends on the willingness of the semiconductor industry in terms of need, infrastructural capital investment, yield and manufacturing cost.

Silver Nanoparticles

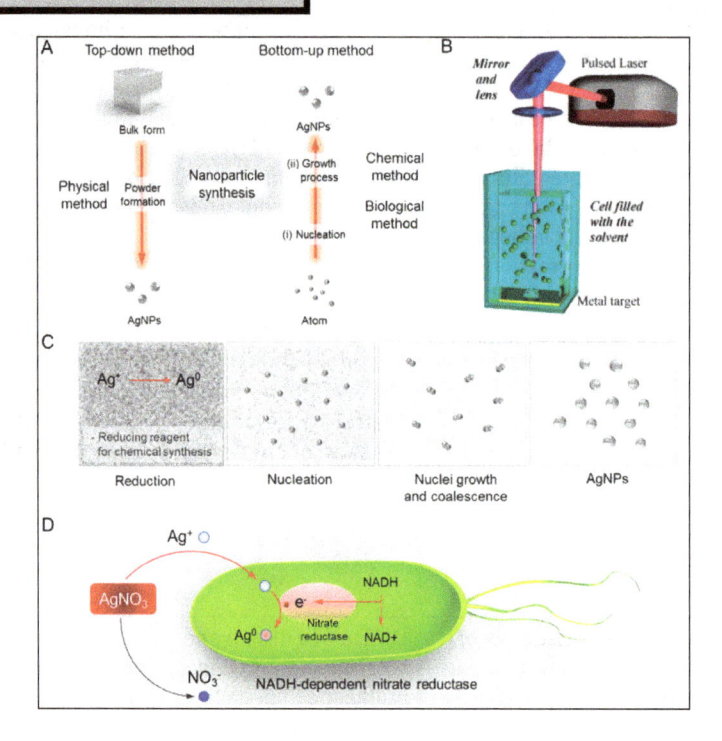

Numerous types of silver nanostructures with distinctive properties have been used in various biomedical fields. In particular, silver nanomaterials of varying sizes and shapes have been utilized in a broad range of applications and medical equipment, such as electronic devices, paints, coatings, soaps, detergents, bandages, etc. Specific physical, optical, and chemical properties of silver nanomaterials are, therefore, crucial factors in optimizing their use in these applications. In this regard, the following details of the materials are important to consider in their synthesis: surface property, size distribution, apparent morphology, particle composition, dissolution rate (i.e., reactivity in solution and efficiency of ion release), and types of reducing and capping agents used.

Diverse synthesis routes of silver nanoparticles (AgNPs): (A) Top-down and bottom-up methods. (B) Physical synthesis method. (C) Chemical synthesis method. (D) Plausible synthesis mechanisms of green chemistry. The bioreduction is initiated by the electron transfer through nicotinamide adenine dinucleotide (NADH)-dependent reductase as an electron carrier to form NAD$^+$. The resulting electrons are obtained by Ag$^+$ ions which are reduced to elemental AgNPs.

The synthesis methods of metal NPs are mainly divided into top-down and bottom-up approaches as shown in figure. The top-down approach disincorporates bulk materials to generate the required nanostructures, while the bottom-up method assembles single atoms and molecules into larger nanostructures to generate nano-sized materials. Nowadays the synthetic approaches are categorized into physical, chemical, and biological green syntheses. The physical and chemical syntheses tend to be more labor-intensive and hazardous, compared to the biological synthesis of AgNPs which exhibits attractive properties, such as high yield, solubility, and stability. Shape and size-controlled synthesis of AgNPs can be achieved through appropriate selection of energy source, precursor chemicals, reducing and capping agent, as well as concentration and molar ratio of chemicals.

Physical Method

The physical synthesis of AgNP includes the evaporation–condensation approach and the laser ablation technique. Both approaches are able to synthesize large quantities of AgNPs with high purity without the use of chemicals that release toxic substances and jeopardize human health and environment. However, agglomeration is often a great challenge because capping agents are not used. In addition, both approaches consume greater power and require relatively longer duration of synthesis and complex equipment, all of which increase their operating cost.

The evaporation–condensation technique typically uses a gas phase route that utilizes a tube furnace to synthesize nanospheres at atmospheric pressure. Various nanospheres, using numerous materials, such as Au, Ag, and PbS, have been synthesized by this technique. The center of the tube furnace contains a vessel carrying a base metal source which is evaporated into the carrier gas, allowing the final synthesis of NPs. The size, shape, and yield of the NPs can be controlled by changing the design of reaction facilities. Nevertheless, the synthesis of AgNPs by evaporation–condensation through the tube furnace has numerous drawbacks. The tube furnace occupies a large space, consumes high energy elevating the surrounding temperature of the metal source, and requires a longer duration to maintain its thermal stability. To overcome these disadvantages, researchers demonstrated that a ceramic heater can be utilized efficiently in the synthesis of AgNPs with high concentration.

Another approach in physical synthesis is through laser ablation. The AgNPs can be synthesized by laser ablation of a bulk metal source placed in a liquid environment as shown in figure B. After irradiating with a pulsed laser, the liquid environment only contains the AgNPs of the base metal source, cleared from other ions, compounds or reducing agents. Various parameters, such as laser power, duration of irradiation, type of base metal source, and property of liquid media, influence the characteristics of the metal NPs formed. Unlike chemical synthesis, the synthesis of NPs by laser ablation is pure and uncontaminated, as this method uses mild surfactants in the solvent without involving any other chemical reagents.

Chemical/Photochemical Methods

Chemical synthesis methods have been commonly applied in the synthesis of metallic NPs as a colloidal dispersion in aqueous solution or organic solvent by reducing their metal salts. Various metallic salts are used to fabricate corresponding metal nanospheres, such as gold, silver, iron, zinc oxide, copper, palladium, platinum, etc. In addition, reducing and capping agents can easily be changed or modified to achieve desired characteristics of AgNPs in terms of size distribution, shape, and dispersion rate. The AgNPs are chemically synthesized mainly through the Brust–Schiffrin synthesis (BSS) or the Turkevich method. The strength and type of reducing agents and stabilizers should be taken into consideration in synthesizing metal NPs of a specific shape, size, and with various optical properties. More importantly, as stabilizing agents are typically used to avoid aggregation of these NPs, the following factors need to be considered for the safety and effectiveness of the method: choice of solvent medium; use of environment-friendly reducing agent; and selection of relatively non-toxic substances.

Nucleation and growth of NPs are governed by various reaction parameters, including reaction temperature, pH, concentration, type of precursor, reducing and stabilizing agents, and molar ratio of surfactant/stabilizer and precursor. The chemical reduction of these metal salts can be accomplished by various chemical reductants, including glucose ($C_6H_{12}O_6$), hydrazine (N_2H_4), hydrazine hydrate, ascorbate ($C_6H_7NaO_6$), ethylene glycol ($C_2H_6O_2$), N-dimethylformamide (DMF), hydrogen, dextrose, ascorbate, citrate (Turkevich method), and sodium borohydride (BSS method). Brust and co-workers have invented the most widely used synthesis method in producing thiol-stabilized AuNPs and AgNPs. Silver ion (Ag^+) is reduced in aqueous solution, receiving an electron from a reducing agent to switch from a positive valence into a zero-valent state (Ag^0), followed by nucleation and growth. This leads to coarse agglomeration into oligomeric clusters to yield colloidal AgNPs. Research using a strong reductant (i.e., borohydride) has demonstrated the synthesis of small monodispersed colloids, but it was found to be difficult to control the generation of larger-sized AgNPs. Utilizing a weaker reductant, such as citrate, resulted in a slower reduction rate, which was more conducive to controlling the shape and size distribution of NPs.

Stabilizing dispersive NPs during a course of AgNP synthesis is critical. The most common strategy is to use stabilizing agents that can be absorbed onto the surface of AgNPs, avoiding their agglomeration. To stabilize and to avoid agglomeration and oxidation of NPs, capping agents/surfactants can be used, such as chitosan, oleylamine gluconic acid, cellulose or polymers, such as poly N-vinyl-2-pyrrolidone (PVP), polyethylene glycol (PEG), polymethacrylic acid (PMAA) and polymethylmethacrylate (PMMA). Stabilization via capping agents can be achieved either through electrostatic or steric repulsion. For instance, electrostatic stabilization is usually achieved through anionic species, such as citrate, halides, carboxylates or polyoxoanions that adsorb or interact with AgNPs to impart a negative charge on the surface of AgNPs. Therefore, the surface charge of AgNPs can be controlled by coating the particles with citrate ions to provide a strong negative charge. Compared to using citrate ions, using branched polyethyleneimine (PEI) creates an amine-functionalized surface with a highly positive charge. Other capping agents also provide additional functionality. Polyethylene glycol (PEG)-coated nanoparticles exhibit good stability in highly concentrated salt solutions, while lipoic acid-coated particles with carboxyl groups can be used for bioconjugation.

On the other hand, steric stabilization can be achieved by the interaction of NPs with bulky groups, such as organic polymers and alkylammonium cation that prevent aggregation through steric repulsion. For instance, researchers described a Brust synthesis-modified procedure for dodecanethiol-capped AgNPs, wherein dodecanethiol could bind onto the surface of nanoparticles and exhibited high solubility without their aggregation in aqueous solution. A phase transfer of a Au3+ complex can be carried out from aqueous to organic solution in a two-phase liquid–liquid system, then the complex can be reduced with sodium borohydride ($NaBH_4$) along with dodecanethiol as a stabilization agent. The authors demonstrated that small alterations in parameters can lead to dramatic modifications in the structure, average size, and size distribution of the nanoparticles as well as their stability and self-assembly patterns.

The surface of AgNPs conjugated with biomolecules, such as DNA probes, peptides or antibodies, can be used as a target for specific cells or cellular components. Attaching biomolecules to AgNPs can be achieved, for instance, by physisorption onto the surface of NPs or through covalent coupling by ethyl(dimethylaminopropyl) carbodiimide (EDC) to link free amines on antibodies to carboxyl groups. The photochemical synthesis method also offers a reasonable potential for the synthesis of shape- and size-controlled AgNPs although multiple synthesis steps may be required. Ag nanoprisms can be synthesized by irradiating Ag seed solution with a light at a selected wavelength.

Commonly, the synthesis of bipyramids, nanodiscs, nanorods, and nano-decahedron involves a two-step process. Ag seeds prepared in the first step are subsequently grown in the second step by using an appropriate growth solution, by selecting a specific wavelength of light for irradiation, or by adjusting the duration of microwave irradiation. To synthesize distinctively shaped AgNPs, selective adsorption of surfactants/stabilizers to specific crystal facets needs to be controlled, since surfactants/stabilizers can guide growth along a specific crystal axis, generating varied shapes of AgNPs. The absorbance spectra of AgNPs have been reported to reflect changes in the shape of AgNPs. Such changes in UV–Vis–NIR spectra were illustrated during the photochemical synthesis of Ag nanoprisms grown by illuminating small silver NP seeds (λ_{max} of 397 nm) with low intensity LED. As the seeds were converted to nanoprisms, the peak wavelength at 397 nm decreased over time, and new peaks appeared at 1330 nm and 890 nm, representing a localized surface plasmon resonance (LSPR) of the nanoprisms. For instance, a research group has investigated photo-induced conversion of spherical AgNPs to triangular prisms. Spontaneous oxidative dissolution of small Ag particles enabled the production of Ag^+ ions that could subsequently be reduced on the surface of Ag particles by citrate under visible light irradiation.

Green Chemistry

Recently, the biogenic (green chemistry) metal NP synthesis method that employs biological entities, such as microorganisms and plant extracts, has been suggested as a valuable alternative to other synthesis routes. It is known that microorganisms, such as bacteria and fungi, play a vital role in remediation of toxic materials by reducing metal ions. Quite a few bacteria have shown the potential to synthesize AgNPs intracellularly, wherein intracellular components serve as both reducing and stabilizing agents. The green synthesis of AgNPs with naturally occurring reducing agents could be a promising method to replace more complex physiochemical syntheses since the green synthesis is free from toxic chemicals and hazardous byproducts and instead involves natural capping agents for the stabilization of AgNPs.

A plausible mechanism of AgNP formation by the green synthesis was explored in the biological system of a fungus, Verticillium species. The main hypothesis was that AgNPs are formed underneath the surface of the cell wall, not in the aqueous solution. Ag^+ ions are trapped on the surface of the fugal cells due to the electrostatic interaction between Ag^+ ions and negatively-charged carboxylate groups of the enzyme. Then, as intracellular reduction of Ag^+ ions occurs in the cell wall, Ag nuclei are formed, which subsequently expand by further reduction of Ag^+ ions. The result of transmission electron microscopy (TEM) analysis indicated that AgNPs were formed in cytoplasmic space due to the bioreduction of the Ag^+ ions, yielding a particle size of 25 ± 12 nm in diameter. Interestingly, the fungal cells continued to proliferate after the biosynthesis of AgNPs.

Bacteria commonly use nitrate as a major source of nitrogen, whereby nitrate is converted to nitrite by nitrate reductase, utilizing the reducing power of a reduced form of nicotinamide adenine dinucleotide (NADH). Bacterial metabolic processes of utilizing nitrate, namely reducing nitrate to nitrile and ammonium, could be exploited in bioreduction of Ag^+ ions by an intracellular electron donor. In fact, the utilization of nitrate reductase as a reducing agent is found to play a key role in the bioreduction of Ag^+ ions. For instance, various researches have demonstrated a rationale of an in vitro enzymatic strategy for the synthesis of AgNPs, based on α-NADPH-dependent nitrate reductase and phytochelatin. Nitrate reductase purified from a fungus, Fusarium oxysporum, was used in vitro in the presence of a co-factor, α-NADPH. The process of AgNPs formation required the reduction of α-NADPH to α-NADP$^+$. Hydroxyquinoline probably acted as an electron shuttle, transferring electrons generated during the reduction of nitrate to allow conversion of Ag^{2+} ions to Ag. As the Ag^+ ions were reduced in the presence of nitrate reductase, a stable silver hydrosol (10–25 nm) was formed and subsequently stabilized by capping peptide.

Similarly, AgNPs have been synthesized in various shapes using naturally occurring reducing agents (i.e., supernatants) in Bacillus species. In Bacillus licheniformis, it was demonstrated that electrons released from NADH were able to drive the reduction of

Ag⁺ ions to Ag⁰ and led to the formation of AgNPs. Researchers also showed the synthesis of AgNPs by reductase enzymes secreted from a fungus, Aspergillus terreus, based on a similar NADH-mediated mechanism. The synthesized AgNPs were polydispersed nanospheres ranging from 1 to 20 nm in diameter and exhibited antimicrobial potential to various pathogenic bacteria and fungi. In another example, Pseudomonas stuzeri isolated from a silver mine was used for the synthesis of AgNPs in aqueous $AgNO_3$. The synthesized AgNPs exhibited a well-defined size and distinct morphology within the periplasmic space of the bacteria.

Characterization and Property of AgNPs

Plasmonic Properties

In many applications, surface chemistry, morphology, and optical properties associated with each NP variant require a careful selection to acquire the desired functionality of nanomaterials. In particular, corresponding reaction conditions during the synthesis of silver nanomaterials can be tuned to produce colloidal AgNPs with various morphologies, including monodisperse nanospheres, triangular nanoprisms, nanoplates, nanocubes, nanowires, and nanorods. Nowadays, since the most commonly used Ag and Au nanospheres are isotropic, they are widely utilized nanostructures for nanoantenna, capitalizing the LSPR phenomena caused by the collective oscillation of electrons in a specific vibrational mode at the conduction band near the particle surface in response to light.

Representative images of electron microscopy of synthesized Ag nanostructures, demonstrating that diverse sizes and morphologies are made possible by controlling the reaction chemistry. (A) Silver nanosphere, (B) Silver necklaces, (C) Silver nanobars, (D) Silver nanocubes, (E) Silver nanoprism, (F) Silver bipyramids, (G) Silver nanostar, (H) Silver nanowire, (I) Silver nanoparticle embedded silica particle.

The optical properties can be varied by changing the composition, size, and shape of

NPs which can affect the collective oscillation of free electrons in metallic NPs at their LSPR wavelengths when irradiated with resonant light over most visible and near-infrared regions. Endowed with the tunable optical response, the NPs can be utilized as highly bright reporter molecules, efficient thermal absorbers, and nanoscale antenna, all through amplifying the strength of a local electromagnetic field to detect changes in the environment. The shape of silver nanoprisms has a specific peak wavelength that ranges from 400 to 850 nm as a surface plasmon resonance (SPR) band as shown in figure. The SPR band or absorption spectra for nanoprisms can be measured by the UV–VIS spectroscopy, whereby the λ_{max} reflects an alteration in the size, shape, and the scattering color of AgNPs. The optical properties of AgNPs have been of particular interest due to the strong coupling of AgNPs to specific wavelengths of incident light. Ag nanospheres are known to have rather short LSPR wavelengths in the violet and blue regions of the visible spectrum.

AgNPs can be utilized in bio-sensing by single nanoparticle spectroscopy, such as dark-field microscopy. Alivisatos and his co-worker described 'plasmon rulers' to monitor distances between two distinct nanoparticles. The distance can be determined by plasmonic coupling of two nanospheres modified at two ends of a single-stranded DNA (ssDNA) probe with biotin on one end and streptavidin on the other end. The authors demonstrated the plasmonic coupling between single pairs of silver and gold nanoparticles to measure the DNA length and tracked the hybridization kinetics over 3000 s. The plasmonic coupling between two distinct nanoparticles led to more pronounced spectral changes based on the dimerization of single nanoparticles. For example, DNA hybridization was responsible for the observed blue-shift in the spectra via an increase in steric repulsion or for the observed drastic red-shift via aggregation, such as DNA wrapping around DNA-binding dendrimers.

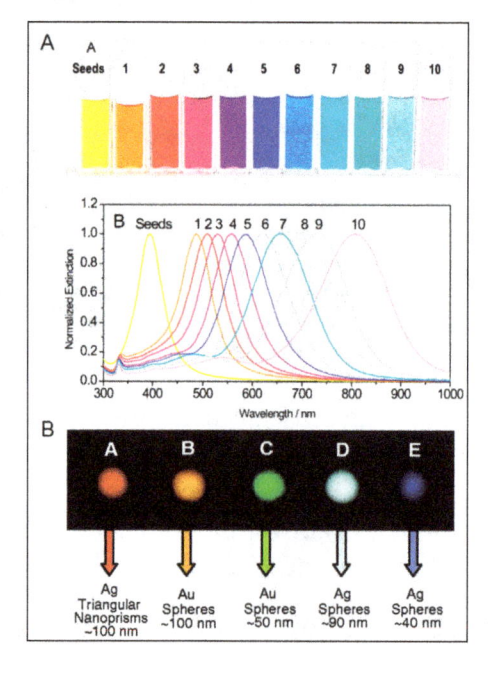

Figure shows (A) Photograph of silver nanoprisms (top) and corresponding optical spectra changes of nanoprisms (bottom). Control on the edge-length of nanoprisms allows the plasmon resonance to be tuned across the visible and near-infrared portions of the spectrum. (B) Dark field microscopy images of (left to right) 100 nm diameter silver triangular nanoprism, 90 nm diameter silver nanosphere, and 40 nm diameter silver nanosphere, illustrating the ability to tune the scattering color of silver nanoparticle labels based on size and shape.

In addition, the distance between the AuNPs was adjusted by controlling the length of ssDNA and by changing the ionic strength of the buffer. The maximum plasmon resonance (LSPR) shifted to the red region at high salt concentrations (0.1 M NaCl), indicating a decreased distance between the two AuNPs due to the reduced electrostatic repulsion of the particles at high ionic strength environments. Conversely, low salt concentrations (0.005 M NaCl) increased electrostatic repulsions and led to the blue shift of the maximum LSPR. Along with these results, the hybridization of complementary DNA also resulted in a significant blue shift, which is expected considering that the structural property of double-stranded DNA (dsDNA) shows greater stiffness than ssDNA, and hence allowing it to repulse two AuNPs.

Several reports have demonstrated that AgNPs absorb electromagnetic radiation in the visible range from 380 to 450 nm, which is known as the excitation of LSPR. The optical properties of AgNPs of different sizes by gallic acid using biological synthesis methods were characterized by Park and colleagues. The authors demonstrated that spherical AgNPs of 7 nm have SPR at 410 nm, while those of 29 nm have 425 nm. In addition, 89 nm-sized AgNPs exhibited a wider band with a maximum resonance at 490 nm. It was noticed that the width of the SPR band was related to the size distributions of NPs. Various Researchers have investigated the dependence in the sensitivity of SPR responses (frequency and bandwidth) that enabled NPs to recognize the changes in their surrounding environment. They also demonstrated how the optical scattering of Au or Ag nanorods with diverse sizes and shapes can affect total extinction. Greater enhancement in the magnitude and sharpness of the plasmon resonance band was observed in nanorods with higher Ag concentration, which could contribute to superior sensing resolution even with a similar plasmon response. As such, Ag nanorods have an additional advantage as better scatterers when compared to Au nanorods.

Chemical Cytotoxicity

One of the current issues in AgNP-based nanomedicine involves nanotoxicity and environmental impact of AgNPs on a nanometer scale. To predict the potential cytotoxic effect of AgNPs, it is necessary to investigate chemical transformation that occurs with AgNPs travelling through the intracellular environment. The use of AgNPs based on their chemical cytotoxic property has received much attention as potent anticancer or antibacterial agents. Despite various hypotheses available, the mechanisms of the antibacterial properties of AgNPs so far have not been established clearly. Based current literature, the proposed cytotoxic mechanisms can be summarized as follows: (i) adhesion of AgNPs onto the membrane surface of microbial cells, modifying the lipid bilayer or increasing the membrane permeability; (ii) intracellular penetration of AgNPs; (iii) AgNP-induced cellular toxicity triggered by the generation of reactive oxygen species (ROS) and free radicals, damaging the intracellular micro-organelles (i.e., mitochondria, ribosomes, and vacuoles) and biomolecules including DNA, protein, and lipids; and (iv) modulation of intracellular signal transduction pathways towards apoptosis. Critical parameters, such as ion release, surface area, surface charge, concentration and colloidal state, can all influence the cytotoxic properties of AgNPs.

The main mechanism of AgNPs regarding their antimicrobial activity can be simplified to their high surface area in releasing silver ions. AgNPs in an aqueous environment are oxidized in the presence of oxygen and protons, releasing Ag^+ ions as the particle surface dissolves. The release rate of the Ag^+ ions depends on a number of factors including the size and shape of NPs, capping agent, and colloidal state. For example, it is well known that antibacterial activity is enhanced with the release of Ag^+ ions from AgNPs onto the bacterial cells. In particular, smaller or anisotropic AgNPs with a larger surface area showed more toxicity and exhibited a faster ion release rate due to high surface energy originating from highly curved or strained shapes of NPs. The small-sized AgNPs also exhibited a superior release rate of silver ion particularly into the Gram-negative bacteria. The shape and higher temperature of AgNPs equally caused a greater degree of toxicity and accelerated the rate of ion release by more effective dissolution. Furthermore, the cytotoxic effect of AgNPs arises in a similar concentration range for both bacteria and human cells. Therefore, a higher Ag ion concentration, a faster release rate of the Ag ions, and a larger surface area of AgNP should be considered for the enhanced antimicrobial treatment in clinical medicine.

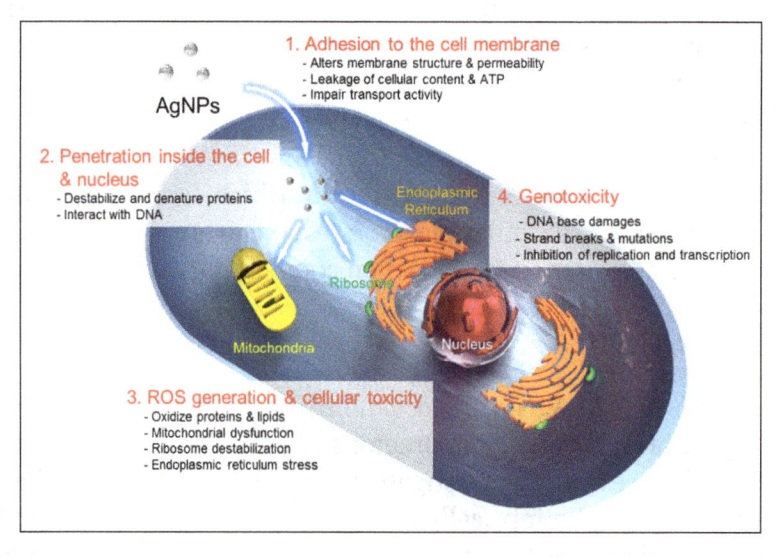

Moreover, the presence of chlorine, thiols, sulfur, and oxygen was shown to strongly impact the rate of silver ion release. Silver ions can interact with thiol groups in critical bacterial enzymes and proteins, and subsequently damage cellular respiration, resulting in cell death. The generation of ROS and free-radicals is another mechanism of AgNPs causing a cell-death process. The potent cytotoxic activity of AgNPs, such as antibacterial, antifungal, and antiviral action, is mainly due to their ability to produce ROS and free radical species, such as superoxide anion (O_2^-), hydrogen peroxide (H_2O_2), hydroxyl radical (OH), hypochlorous acid (HOCl), and singlet oxygen. When in contact with bacteria, the free radicals have the ability to generate pores on the cell wall, which can ultimately lead to cell death. AgNPs can also anchor to the surface of the bacterial cell wall and penetrate it to cause structural changes to the membrane or increase its permeability, all of which trigger cells to die.

The four main routes of cytotoxic mechanism of AgNPs: (1) AgNPs adhere to the surface of a cell, damaging its membrane and altering the transport activity; (2) AgNPs and Ag ions penetrate inside the cell and interact with numerous cellular organelles and biomolecules, which can affect corresponding cellular function; (3) AgNPs and Ag ions participate in the generation of reactive oxygen species (ROS) inside the cell leading to a cell damage and; (4) AgNPs and Ag ions induce the genotoxicity.

Interestingly, the strength of the antibacterial property of AgNPs is correlated with different types of bacterial species, such as Gram-positive and -negative bacteria. This is because these species differ in the architecture, thickness, and composition of their cell wall. It is known that Escherichia coli (E. coli) which is Gram-negative bacteria are more susceptible to Ag^+ ions than Gram-positive Staphylococcus aureus (S. aureus). The reason for different susceptibility lies on the peptidoglycan which is a key component of the bacterial cell membrane.

The cell wall in Gram-positive bacteria is composed of a negatively-charged peptidoglycan layer with approximately 30 nm in thickness, whereas Gram-negative bacteria have a peptidoglycan layer of only 3 to 4 nm. These structural differences, including the thickness and composition of the cell wall, explain why Gram-positive S aureus is less sensitive to AgNPs, and Gram-negative E.coli displays substantial inhibition even at a low concentration of AgNPs. Several researchers investigated silver and curcumin NPs against both Gram-positive and Gram-negative bacteria, and the NPs in 100 µg/mL concentration were able to distort matured bacterial biofilms. The sustained anti-bacterial effects of this formulation can be utilized in antimicrobial treatment. Researcher investigated the anti-fungal effects of AgNPs and copper nanospheres against wood-rotting fungi. The AgNP treatment required a very low mass of NPs and exhibited high efficiency against Tinea versicolor (T. versicolor) fungi in comparison to Poria placenta (P. placenta) fungi, showing differing anti-fungal effects of AgNPs against white and brown-rot fungi, respectively.

Alloy with other Metals

Alloy NPs exhibit specific properties that are different from their individual NPs. They can be created directly by combining different metallic nanocrystals (NCs) in specific numbers and arrangements. As strong electronic coupling exists between two metals, the bimetallic nanocrystals show more enhanced catalytic, electronic and optical properties compared to monometallic nanocrystals. The properties of alloy NPs are defined by their internal configuration (i.e., arrangement of constituent atoms) and external structures, such as shapes and sizes. The LSPR wavelength of nanocrystals formed from Ag–Au alloys can be tuned by varying the Au:Ag ratio and, therefore, can increase gradually with an increase in the percentage of Au in the alloy nanocrystal. However, metallic atoms are easily bonded in a non-specific manner, lacking the directionality of covalent bonds and equivalence of molecules. The synthesis methods for alloy NPs can be divided into two categories: (i) successive/sequential reduction method and (ii) co-reduction/simultaneous reduction method with metal precursors.

Sequential reduction without protective agents is driven thermodynamically and causes the formation of core–shell NPs or other types of hetero-nanostructures. The sequential reduction method involves subsequent seed-mediated growth of NPs with metal precursors and reducing agents over time. The bimetallic colloids with different metals, such as Ag and Au, can be synthesized in several different ways, resulting in Ag-coated Au or Au-coated colloidal particles. The synthesis can be done simply by reducing one metal salt on already-formed counterpart metal NPs—e.g., to synthesize Ag-coated Au colloids, chemically reduce silver salt on the AuNPs. This seed-mediated synthesis method for core–shell and intermetallic structures is widely used in well-defined bimetallic NPs, due to its capability to regulate the size, shape and composition of the final compound. Reducing agents also play a vital role in controlling the size distribution. The co-reduction method with different metal precursors to zero-valent atoms has made bimetallic colloids readily accessible. The key advantages lie in the simplicity and versatility of the technique. Using this method, several types of Ag and Au bimetallic core–shell colloids with various shapes have been produced. Bimetallic colloids with gradient metal distribution or with a layered structure are one of the most interesting and promising methods in catalytic applications. In the co-reduction method, however, composition uniformity is a major drawback due to the high prevalence of sequential reduction.

Applications of Silver Nanoparticles

Recently, AgNP has been widely utilized in various subfields of nanomedicine including nanoelectronics, diagnostics, molecular imaging, and biomedicine. These interesting applications are based on utilizing an enhanced electromagnetic field on and near the surface of AgNPs. At the plasmon resonant wavelength, AgNPs act as nanoscale antennas, increasing the intensity of a local electromagnetic field. One spectroscopic technique that benefits from the enhanced electromagnetic field is the Raman spectroscopy, where molecules can be identified by their unique vibrational modes. However, intrinsic Raman scattering of photons from molecules is weak and requires longer measurement duration to obtain a Raman spectrum.

Therefore, surface-enhanced Raman scattering (SERS) from molecules near the surface of plasmonic nanoantenna offers great amplification of Raman signals. Typically, SERS detection involves adsorption of molecules on Ag or Au nanoparticle aggregates or solid substrates with plasmonic nanostructures. Strong field enhancement is generated in the nanogaps or interstices known as hot spots within interacting plasmonic nanostructures. The SERS effect can be used to detect critical proteins and biomolecules, such as early cancer biomarkers or drug levels in blood and other body fluids. Up until recently, the SERS effect with hot spots has been the main focus in numerous experimental and theoretical studies, which can enhance the Raman scattering to the factor of 10^8 to 10^{12}, allowing the detection of even a single molecule.

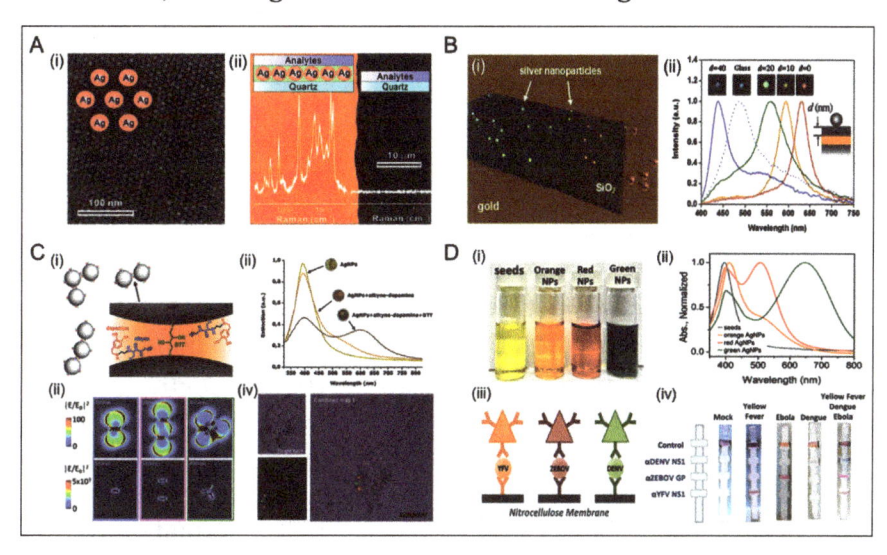

Plasmonic AgNPs for plasmonic nanoantennas and diagnostics: (A) Single-layer AgNP surface-enhanced Raman scattering (SERS) film for a large-scale hot spot. (i) Scanning electron microscopy (SEM) image of a superlattice of 6 nm. AgNPs were used as a homogeneous single-molecule SERS substrate. Illustration shows an interparticle gap for hot spots, which is regulated by the length of a thiolate chain. (ii) Two Raman spectra of single-layered SERS film (left) and quartz surface (right). The enhancement factor was estimated to be larger than 1.2×10^7. (B) Metal-film induced plasmon resonance tuning of AgNPs. (i) Schematic illustration of optical scattering spectra of AgNPs on different substrates. (ii) Single AgNP spectra of AgNPs on a silica spacer layer of varying thickness d (nm) on a glass substrate with a 50 nm gold film. The inset is a dark-field image of AgNPs with the corresponding color. The dotted lines represent single particle spectra of AgNPs on a plain glass substrate. (C) SERS-based intracellular imaging using alkyne-AgNPs nanoprobes. (i) The structure of colloidal alkyne-AgNP clusters with nano-sized interparticle gaps. (ii) Extinction spectra of the alkyne-AgNPs nanoprobe. The resonance peaks at 400 nm shifted around 520 nm after metal functionalization. (iii) Computational simulation of the far- and near-field optical responses. Intensity distributions of the single particle mode (upper-panels) and the dimer mode (bottom-panels) (iv) Intracellular Raman imaging of a AgNP nanoprobe within the cytoplasmic space of fibroblast. Distinguishable hot spots were highlighted by color-dots related to Raman intensity of the akyne 2045 cm^{-1} band. (D) Multiplexed detection with a tunable wavelength of AgNPs. (i) Different colors of AgNPs during a stepwise growth. (ii) Corresponding absorption spectra with varying sizes of AgNPs, such as 30, 41, and 47 nm. (iii) Individual testing of Yellow Fever virus (YFV) NS1 protein, Zaire Ebola virus (ZEBOV) glycoprotein (GP), and Dengue virus (DENV) NS protein using AgNPs. Orange, red, and green AgNPs were conjugated with monoclonal antibodies specific to YFV NS1, ZEBOV GP, and DENV NS, respectively. (iv) Multiplexed detection using different AgNPs-based lateral flow assays.

Numerous approaches have been made to utilize the plasmonic property of AgNPs. For instance, techniques to control the distance and spacing of hot spots are essential in quantitative SERS covering large areas as shown in figure A, B. Strong enhancement of single hot spots may lead to a false representation of a sample when the signal is mainly determined by a few detection sites. Nanoparticle superlattices have demonstrated a potential to counterbalance the homogeneous distribution in sensing hot-spot bands and to enhance the detection performance of the sensor. Researchers have explained a SERS substrate via graphene–AgNPs heterojunction which improved Raman signals. With increasing the density of AgNPs, Raman scattering of graphene-veiled AgNPs heterojunction substrate was significantly enhanced by approximately 67 folds compared to R6G analyte. The cooperative synergy generated by the coupling of graphene and deposited AgNPs can be utilized to create a strong electromagnetic hot spot for an optical sensing platform. Another example is a star-shaped Au/AgNPs SERS substrate on Ge (5 nm)/Ag (25 nm)/GE (75 nm)/glass (germanium–silver multilayers) which was employed via near-infrared (NIR) SERS operation. The hybrid SERS substrate was operated at a 1064 nm excitation and exhibited 30% higher Raman intensity.

AgNPs can be utilized as highly sensitive NP probes for targeting and imaging of small molecules, DNA, proteins, cells tissue, and even tumor in vivo. AgNPs with stronger and sharper plasmon resonance have been widely used in imaging systems, particularly for cellular imaging with contrast agents functionalized to AgNPs via surface modification. For example, AgNP-embedded nanoshell structure can be used in cancer imaging and photothermal therapy to explore the location of cancer cells by absorbing light and destroy them via photothermal effect. Researchers described NIR-sensitive SERS nanoprobes for an in vivo multiplex molecular imaging to detect aromatic compounds. The NIR SERS probes with plasmonic Au/Ag hollow-shell were assembled onto silica nanospheres which exhibited a red-shift of plasmonic extinction band in the NIR optical window region (700–900 nm). The signals from the NIR-SERS nanoprobe for single particle detection exhibited a detectable signal from animal tissues that were 8 mm deep. Researcher showed Ag-embedded SERS nanoprobes, called M-SERS dot, which have a Raman signature for imaging of target cancer cells as well as strong magnetic properties for identifying desirable cells. AgNP-embedded magnetic nanoparticles (MNPs), which consisted of a magnetic core (18 nm) and a silica shell (16 nm thick) decorated with AgNPs on the surface, were prepared.

The M-SERS dots exhibited strong SERS signals originating from diverse encoding materials, such as AgNPs and Raman-labels. The Ag-embedded M-SERS dots with highly sensitive SERS signals enabled targeting, isolation, and imaging of cancer cells. To investigate their specific targeting and sorting abilities, M-SERS conjugated with targeting antibodies were added into multiple cell population, and subsequently, the targeted cancer cells could easily be isolated by an external magnetic field. Various studies have described a multilayered core–shell nanoprobe with Ag-embedded silica nanostructure for a SERS-based chemical sensor. The multi-layered nanoprobe consisted of a silica core coated with Raman label, silica shell, and AgNPs. The embedded inner AgNP and Raman label compound in the nanoprobe served as an internal standard for calibrating SERS signals, while the outer AgNPs were utilized as a sensing site for analyte detection. These chemical sensors based on the ratiometric analysis ($I_{Analyte}/I_{Internal\ standard}$) could be applied to various SERS probes for quantitative detection of a wide variety of targets.

Diagnostics with Tunable Wavelength

AgNPs can absorb and scatter light with extraordinary efficiency. A large scattering cross-section of the nanospheres allows for an individual AgNP to be imaged under a dark-field microscopy or hyperspectral imaging systems. AgNPs have been intensively utilized in several applications, including diagnosis and bioimaging of cancer cells. Furthermore, AgNPs have been utilized for the detection of p53 in carcinoma cells. Researchers have demonstrated that nanostructures comprising silver cores and a dense layer of Y_2O_3:Er separated by a silica shell is an excellent system model to investigate the interaction between upconversion materials and metals on a nanoscale.

Nanoparticles are also potentially promising as fluorescent labels for (single particle) imaging experiments or bioassays, which require low background or tissue penetrating wavelengths. Optical properties of AgNPs can be utilized for multiplexed point-of-care (POC) diagnostics using their size-tunable absorption spectra. Multicolored AgNPs-based multiplexed lateral flow assay (LFA) for multiple pathogen detections. Multiplexed rapid LFA diagnostics has the ability to discriminate among multiple pathogens, thereby facilitating effective investigations for diagnosis. Specifically, triangular plate-shaped AgNPs with varying sizes, such as 30 nm, 41 nm and 47 nm, have narrow absorbance that are tunable through the visible spectrum, resulting in an easily distinguishable color.

The multicolored AgNPs were conjugated with antibodies to recognize dengue virus (DENV) NS protein, Yellow Fever virus (YFV) NS1 protein, and Zaire Ebola virus (ZEBOV) glycoprotein (GP). The limit of detection (LOD) for the biomarkers of each virus was 150 ng/mL in a single channel. Another example is a colorimetric lead detection using AgNPs described by Balakumar and co-workers. The synthesized AgNPs exhibited high sensitivity for the detection of as low as 5.2 nM of Pb^{2+} in the range of 50 to 800 nM and also showed selective recognition even in the presence of interfering metal ions. This approach can be used for a rapid and cost-effective detection of saturnism (lead poisoning) in a water sample.

Surface-enhanced Fluorescence

Surfaces of metallic nanoparticles can alter the free-space condition of fluorescence with spectral properties which can result in dramatic spectral changes as shown in figure (A) below. These interactions between metal surface and fluorophore have been termed variously as surface-enhanced fluorescence (SEF), metal-enhanced fluorescence (MEF) or radiative decay engineering. The metallic surfaces exhibit the following features: fluorophore quenching within short distances (0–5 nm); spatial disparity of incident light (0–15 nm); and changes in the radiative decay rates (1–20 nm). The enhanced field effect can be leveraged to build an interspace with a shorter distance between a fluorophore and the surface of a metal nanostructure composed of Ag or Au, increasing fluorophore emission rate. The enhanced fluorescence can be attributed to two main factors, which are; (i) an enhanced excitation rate by large absorption and scattering cross-section of the plasmonic nanoparticles to incoming light and (ii) a decrease in fluorescence lifetime of the fluorophore that allows an excited state to return to the ground state at a higher frequency. In particular, SEF parameters of the dye/nanoparticle coupled system are related to the distribution of near-field intensity and their distance-dependent decay function. Such effects depend strongly on the overlap of optical properties of the fluorophore and nanosphere and on the physical location of the fluorophore around the particles.

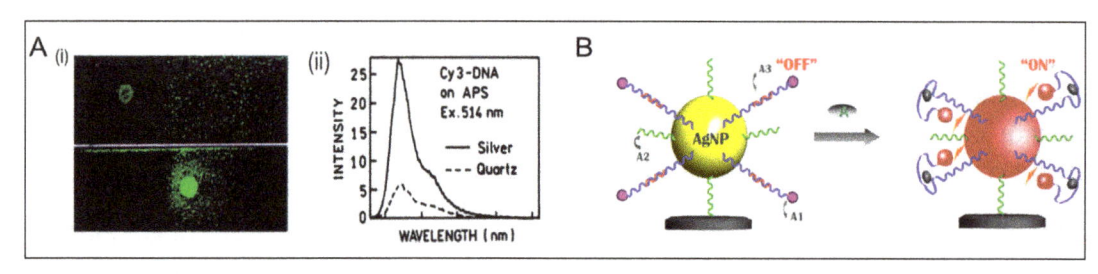

Surface-enhanced fluorescence: (A) Metal-enhanced fluorescence on Ag film. (i) The figure shows fluorescence spots on quartz (top) and silver (bottom) taken through 530 nm long pass filter for Cy3-DNA. (ii) Emission spectra of Cy3-DNA on APS-treated slides, with (solid line) and without silver island films (dotted line). (B) Schematic illustration of an aptamer-based AgNP nanosensor, showing the 'off' state via fluorophore quenching within short distances (left) and 'on' state via turn-on fluorescence signal (right) based on the spacing distance between the Cyanine 3 and the AgNP surface in the detection of adenosine.

Biomedical Application of AgNPs

Owing to their intrinsic cytotoxicity, AgNPs have been broadly used as antibacterial and anticancer agents and for biomedical application in the healthcare industry. The degree of toxicity against cells is determined by the surface charges of AgNPs. A positive surface charge of AgNPs renders them more suitable to stay for a longer duration on the tissue surface or luminal side of the blood vessel, which is a major route for administrating anticancer agents. The intrinsic cytotoxic property of AgNPs has been applied against various types of cancer cells, such as hepatocellular carcinoma, lung and breast cancer, and cervical carcinoma. Small sized AgNPs were more efficient in ROS production. Apart from these cellular mechanisms, AgNPs have also shown anti-angiogenic and anti-proliferative properties. The anti-proliferative property mediated by AgNPs in cancer cells is due to their ability to damage DNA, break chromosome, produce genomic instability, and disrupt calcium (Ca^{2+}) homeostasis which induces apoptosis and causes cytoskeletal instability. The cytoskeletal injury blocks the cell cycle and division, promoting anti-proliferative activity of cancer cells.

References

- Bravais-lattice, chemistry: byjus.com, Retrieved 21, April 2020
- What-are-energy-bands, physics: byjus.com, Retrieved 05, July 2020
- Optical-Properties-of-Nanoparticles-in-Composite-Materials- 560118: diva-portal.org, Retrieved 18, February 2020

3

Quantum Dots

The man-made nanoscale crystals which have optical and electronic properties which differ from larger particles are known as quantum dots. When quantum dots are exposed to UV light, electrons in the quantum dots get excited to a state of higher energy. This chapter closely examines the key concepts of quantum dots to provide an extensive understanding of the subject.

Core Structure of Quantum Dots

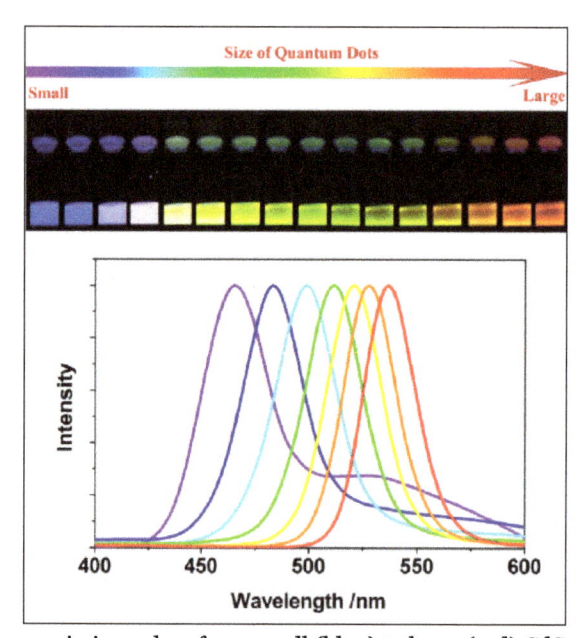

Figure: Top: Sixteen emission colors from small (blue) to large (red) CdSe Qdots excited by a near-ultraviolet lamp; size of Qdots can be from ~1 nm to ~10 nm.
Bottom: Photoluminescence spectra of some of the CdSe Qdots.

Qdots have dimensions and numbers of atoms between the atomic-molecular level and bulk material with a band-gap that depends in a complicated fashion upon a number of factors, including the bond type and strength with the nearest neighbors. For isolated atoms, sharp and narrow luminescent emission peaks are observed. However, a nanoparticle, composed of approximately 100–10000 atoms, exhibits distinct narrow optical line spectra. This is why, Qdots are often described as artificial atoms (δ-function-like DOS). A significant amount of current research is aimed at using the unique optical properties of Qdots in devices, such as light emitting devices (LED), solar cells and biological markers. The most fascinating change of Qdots with particle size <~30 nm is the drastic differences in the optical absorption, exciton energies and electron-hole pair recombination. Use of these Qdot properties requires sufficient control during their synthesis, because their intrinsic properties are determined by different factors, such as size, shape, defect, impurities and crystallinity. The dependence on size arises from (1) changes of the surface-to-volume ratio with size, and from (2) quantum confinement effects (discussed later). Nevertheless, Qdots exhibit different color of emission with change in size. Figure shows change of photoluminescence (PL) emission color with size for CdSe Qdots.

Working of Quantum Dots

Since quantum dots demonstrate size-dependent properties, their behavior can be easily tuned by controlling their size. Analogous to the atoms, quantum dots can be engineered to function as per requirements. For example, an atom can be excited by irradiating it with some energy: that is, an electron from a lower level can be excited to higher energy level. When this excited electron returns back to its original energy level, it emits light having energy equal to the energy difference between the two energy levels involved in the process. The color (or wavelength as well as frequency) of emitted light is characteristic to the atom. For instance, iron appears green when its atoms are excited upon exposure to heat, sodium gives off yellow color. This is attributed to the specific arrangement of energy levels within an atom. Thus, different atoms emit light of different colors or wavelength. This difference comes due to the discretization of energy levels.

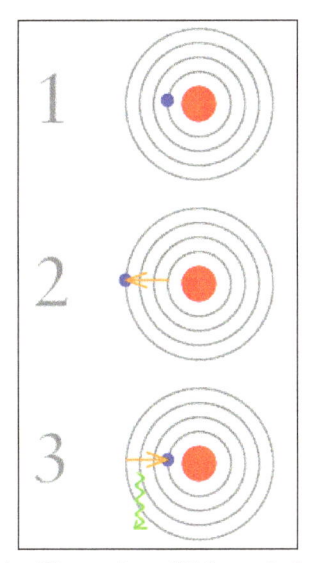

Figure: Schematics illustration of light emission from an atom.

Figure shows schematics of the mechanism behind light emission from an atom. The steps involved can be summarized as follows:

- The atom is in its ground state with no electron present in excited states.

- Light of a specific wavelength is incident onto the atom, which is absorbed by the atom, thereby promoting an electron from lower energy level to a higher energy level.

- The electron returns to its original energy level (where it was present before excitation) by emitting a light photon. The color of the emitted photon or light depends upon the energy difference between the two energy levels involved in the process. As different atoms have different arrangements of their energy levels, dissimilar atoms emit light of varying colors.

A quantum dot also emits light by similar mechanism, because the charge carriers (electrons and hole) entrapped within it has discrete energy levels. However, the arrangement of energy levels within a quantum dot is also influenced by the dimensions of the dot. Thus a quantum dot made from same material can be tuned to emit light of different wavelengths or colors, depending upon the size of the dot. In this regard, bigger quantum dots emit light of longer wavelengths (or smaller frequencies), while smaller quantum dots produce light of shorter wavelength (higher frequency). These large dots give off red light and smaller dots produce blue light, and intermediate dots produce green light. In this way, by adjusting the size of the quantum dot, almost all the colors of the spectrum can be produced.

This can be explained as follows: The band gap of a smaller quantum dot is more than that of a bigger quantum dot. Band gap can be described as the minimum energy required by the atom to promote an electron from the valance band to the conduction band. In this respect, a smaller quantum dot requires more energy to excite the electron. Since the frequency of the emitted light is directly proportional to the energy, small dots produce light having higher frequencies (or shorter wavelength). Similarly, larger dots more closely spaced energy levels, thus they produce lower frequency (or higher wavelength) light.

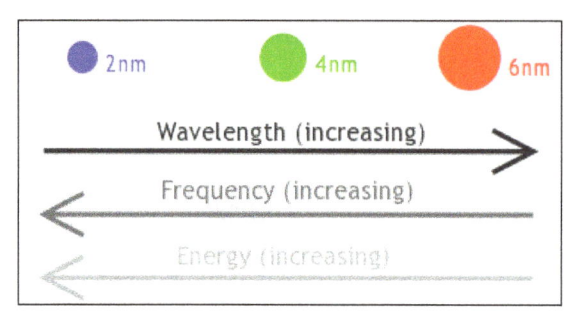

Figure: The size of the quantum dots is in nanometer range. Bigger dots emit light of longer wavelength and lower frequency, such that the color of the emitted light is close to red. Similarly, smaller dots produce shorter wavelength and higher frequency, and the color is towards blue.

Properties of Quantum Dots

Electronic excitation can occur in a semiconductor when a photon with energy equal to or greater than Eg is absorbed. The promoted electron leaves behind a vacancy; the electron deficient region behaves as a positively charged particle and is referred to as a hole. The energy and distribution of the photo generated electrons and holes are described quantum mechanically by their wave functions, probability distributions, and permitted energies. The breadth of the probability distributions of the electrons and holes determine their effective size and are referred to as the electron a_e and hole ah Bohr radii. As for all charged particles, there are Coulombic interactions between the electrons and holes, and the strength of these interactions determine if the electrons and holes can be treated separately or as coulombically bound electron-hole pairs, termed excitons. There is a preferred separation distance between the electron and hole probability distributions in an exciton, and this distance is termed the exciton Bohr radius, a_B.

Luminescence Properties

After excited by an external energy, e.g., photon for photoluminescence, electric field for electroluminescence, primary electron for cathode-luminescence etc., electron and hole possess high energies due to transitions of electron from ground state to an excited state. The energies associated with such optical absorptions are directly determined by the electronic structure of the material. The excited electron and hole may form an exciton, as discussed above. The electron may recombine with the hole and relax to a lower energy state, ultimately reaching the ground state. The excess energy resulting from recombination and relaxation may be either radiative (emits photon) or non-radiative (emits phonons or Auger electrons). Some radiative events from band-edge, defects and non-radiative processes are discussed in brief. inescence etc., electron and hole possess high energies due to transitions of electron from ground state to an excited state. The energies associated with such optical absorptions are directly determined by the electronic structure of the material. The excited electron and hole may form an exciton, as discussed above. The electron may recombine with the hole and relax to a lower energy state, ultimately reaching the ground state. The excess energy resulting from recombination and relaxation may be either radiative (emits photon) or non-radiative (emits phonons or Auger electrons).

Quantum Confinement

Decreasing the size of a semiconductor particle in one, two or all three physical dimensions can change the electronic structure of the material by confining the spatial distribution of the exciton, electron, or even hole wave functions. Since there is a continuum of states for each of the dimensions that do not exhibit quantum confinement, therefore, the density of states depend on the dimensionality of the material, as shown in figure below. The states of a QD become discrete with energies of a few hundred meV between the quantized states, figure below, and there is a similar shift of the band gap to higher energy than the bulk Eg. In a 2D system the states resemble a saw-like quasi-continuum, the energy between these states can be a few hundred meV, and the band gap can also shift a few hundred meV above eg.

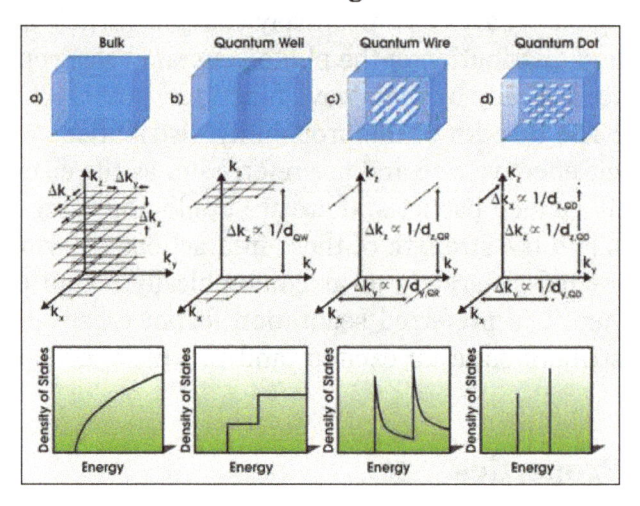

Figure above shows the density of states in energy for a semiconductor as the degree of quantum confinement increases. On the left, a), shows the continuous nature of the bulk material, b) depicts confinement in 1 dimension, c) in two and d) the density of states becomes discrete for confinement in all three dimensions.

The strong, medium, and weak confinement terms mentioned are a result of the specific size in each dimension relative to a_B, a_e, and ah. In the case of strongly-confined systems, the restricted dimension(s) are smaller than both a_e and ah in an intermediate-confinement regime, the size of the confined dimension is smaller than a_e but greater than ah. In a weakly confined system, the restricted dimension(s) are larger than both a_e and ah but small enough that the wavefunction of the exciton is perturbed, and an energetic shift of this state to higher energy results.

The extent of confinement also affects the spatial overlap of the electron and hole wave functions and the Coulombic interactions between them. Bound electron-hole pairs, or excitons, should also be treated as quantum-mechanical systems with wave functions that resemble this simple H-atom system. The photo generated electron-hole interactions are weak in bulk semiconductors, and the binding energies are sufficiently small that the electron-hole pairs are not bound at room temperature. Instead, they dissociate into separate electrons and holes. The dependence of the oscillator strength and the binding energy depends on how many degrees of confinement there are. The energy interactions due to quantum confinement are often large compared to the coulomb interaction term. Furthermore, the dimensionality may not permit the electrons and holes to be stabilized at a preferred length within the coulombic potential, and excitons may not be formed. Such is the case for small QDs, where the electrons and holes are forced to occupy the same volume and the electrons and holes weakly interact and can be considered as independent systems. QDs are novel nanocrystalline semiconductor materials whose electrical as well as optical properties strongly depend upon the size and shape of the dots. The diameter of QDs can vary from ~2 to 10 nm, or on the order of 10-50 atomic lengths. Due to their small size, QDs have very large surface-to-volume ratios. As a result, they have properties somewhere between the individual atoms/molecules and the bulk materials. QDs can be produced either from a single element (e.g., silicon, germanium, and so on) or from more than one element (such as CdSe, CdS, and so on).

Bound Exciton Energy

Coulombic attractions occur between the electron (having negative charge) and hole (having positive charge). The energy of these attractions is directly proportional to Rydberg's energy, and varies inversely with the square of the dielectric constant. This term assumes significance as the semiconductor crystal becomes smaller than the Bohr radius of exciton, and is given by,

$$E_{exciton} = -\frac{1}{\varepsilon_r^2}\frac{\mu}{m_e}Ry = -R_y^*$$

Substituting the values of these energies in the expression for total energy, we obtain,

$$E = E_{bandgap} + E_{confinement} + E_{exciton}$$

$$\therefore E = E_{bandgap} + \frac{h^2\pi^2}{2\mu a^2} - R_y^*$$

Electronic Properties

Small-sized crystals have large electronic band gaps, implying the occurrence of a larger energy difference between energy states. In these cases, larger energy is required to promote the electron to the excited state, and also, more energy is emitted when electrons return to their original energy level. When a semiconductor is incident with a light photon having more energy than its band gap, an electron is excited from the valence band to the conduction band. The excitation of electron to a higher energy state results in the creation of a hole in the valence band. A hole can be simple described as the absence of an electron and it has opposite charge to the electron, i.e., it is positively charged. The electron and hole experience Coulombic attractions and form an exciton.

The exciton, itself, is an electrically neutral quasiparticle, and it can be found in semiconductors, insulators, and also in some liquids. Yakov Frenkel first introduced the concept of exciton in 1931 during a discussion regarding the excitation of atoms in an insulator lattice.

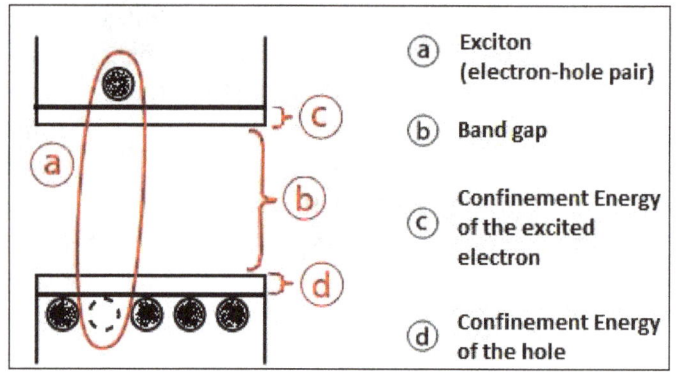

Figure: Generation of exciton from an electron-hole pair.

Types of Emission

Band-edge Emission

The most common radiative relaxation processes in intrinsic semiconductors and insulators are band-edge and near band-edge (exciton) emission. The recombination of an excited electron in the conduction band with a hole in the valence band is called band-edge emission. As noted above, an electron and hole may be bound by a few meV to form an exciton. Therefore, radiative recombination of an exciton leads to near band-edge emission at energies slightly lower than the band-gap. The lowest energy states in Qdots are referred as 1se-1sh (also called exciton state). The full width at half maximum (FWHM) of a room-temperature band-edge emission peak from Qdots varies from 15 to 30 nm depending on the average size of particles. For ZnSe Qdots, however, the luminescence can be tuned by size over the spectral range 390–440 nm with FWHM as narrow as 12.7–16.9 nm.

The optical absorption spectrum reflects the band structure of the materials. While PL from bulk semiconductors is fairly simple and well-understood, and can be explained by parabolic band theory, the PL from Qdots raises several questions. For example, radiative lifetime of 3.2 nm sized CdSe Qdots can be 1 µs at 10K compared to bulk (~1 ns). This was explained by the fact that there were surface states that involved in emission. Band structures of semiconductors are often determined from either absorption spectra or PLE spectra. The study also showed that these two spectra exhibited different characteristics when these spectra were acquired at 15 K. The PLE spectrum was associated with couple of additional peaks along with 1se-1sh. Bawendi et al. assigned these peaks as formally forbidden 1se-1ph and 1se-2sh. It was also observed experimentally that the Stokes shift was size dependent. For a large size CdSe Qdots (5.6 nm), the Stokes shift was found to be 2 meV whereas for a same Qdots of size 1.7nm, the value could be 20 meV. Such a discrepancy was explained, theoretically and experimentally, in terms of increase of distance between optically active state and optically forbidden ground exciton state with decreasing the size of Qdots.

Qdots have a number of advantages over organic dyes in bio applications, e.g., better photostability, wide absorption edges, and narrow, tunable emission. However, they may exhibit a random, intermittent luminescence which is called 'blinking'. In blinking, a Qdot emits lights for a time followed by a dark period.

Defect Emission

Radiative emission from Qdots also comes from localized impurity and/or activator quantum states in the band-gap. Defect states lie inside the bands themselves. Depending on the type of defect or impurity, the state can act as a donor (has excess electrons) or an acceptor (has a deficit of electrons). Electrons or holes are attracted to these sites of deficient or excess local charge due to Coulombic attraction. Similar to the case of excitons, trapped charge on defect/impurity sites can be modeled as a hydrogenic system, where binding energy is reduced by the dielectric constant of the material. These defects states can be categorized into either shallow or deep levels, where shallow level defect states have energies near the conduction band or valence band-edge. In most cases, shallow defect exhibits radiative relaxation at temperatures sufficiently low so that thermal energies (kT) do not excite the carriers out of the defects or traps states. Deep levels, on the other hand, are so long-lived that they typically experience non radiative recombination. Luminescence from these defect levels can be used to identify their energy and their concentration is proportional to the intensity. Both PL spectral distribution and intensity change with changes of the excitation energy due to contributions from different defect energy levels and the band structure of the host. The excitation energy also determines the initial photo excited states in the sample, but this state is short-lived because of thermalization of the photo excited carriers via phonon emission, as discussed above. Relaxation to within kT of the lowest vibrational level of the excited states is usually orders of magnitude faster than the recombination event.

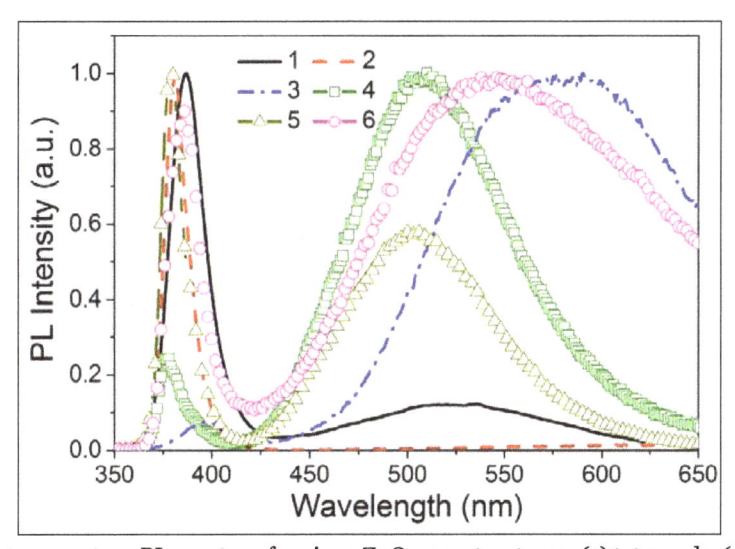

Figure: Room temperature PL spectra of various ZnO nanostructures: (1) tetrapods, (2) needles, (3) nanorods, (4) shells, (5) highly faceted rods, and (6) ribbons/combs.

Defect states are expected at the surface of a Qdot despite the use of various passivation methods, because of the large surface-to-volume ratio, discussed above. The concentration of surface states on the Qdots is a function of the synthesis and passivation processes. These surface states act as traps for charge carriers and excitons, which generally degrade the optical and electrical properties by increasing the rate of non-radiative recombination. However, in some cases, the surface states can also lead to radiative transitions, such as in the case of ZnO nanostructures. Powders of ZnO have a green emission from defects along with a band-edge near UV emission (the band-gap of ZnO is 3.37 eV or 386 nm) at room temperature.

Activator Emission

Luminescence from intentionally incorporated impurities is called extrinsic luminescence. The predominant radiative mechanism in extrinsic luminescence is electron-hole recombination which can occur via transitions from conduction band to acceptor state, donor state to valance band or donor state to acceptor state. In some cases, this mechanism is localized on the activator atom center. In some many cases, the selection rule is relaxed due to mixing of orbitals, such as d-p mixing in a crystal or ligand field where the orbitals are split into hyperfine structures. Therefore, d-d transition is allowed in some cases for transition elements. For Mn^{2+}, the lifetime of the luminescence is in the order of millisecond due to the forbidden d-d transition. Similarly, f-f transition are also often observed for rare earth elements (e.g., Tm^{3+}, Er^{3+}, Tb^{3+}, and Eu^{3+}), although the f levels are largely unaffected by the crystal field of the host due to shielding by the outer s- and p- orbitals. Due this shielding effect, f-f transitions typically have atomic-like sharp peaks in the emission spectra.

The optical properties of doped ZnO Qdots have also been widely investigated. Doping with Er or Mn has been reported to result in preferential orientation of nanorods perpendicular to the substrate. ZnO Qdots have also been doped with rare-earth elements, such as Tb, Ce, Eu and Dy. In the case of Tb-doped ZnO Qdots, emissions from both Tb and defect states were observed. The emission from Tb was found to increase with increasing Tb concentrations, while that from defect states decreased. Eu-related emission was observed from ZnO:Eu nanorods for a suitable excitation wavelength. However, Dy-doped ZnO nanowires exhibited a relatively strong UV emission with a very weak emission from Dy. The effects of doping Mn in ZnO nanoparticles depend strongly on the synthesis conditions. The Mn was found to quench green emission, while others reported either a reduction in both UV and defect emissions or a blue shift and increase in UV peak intensity. Very similar spectra from ZnO and Mn-doped ZnO were observed after annealing at 800 °C. Other dopants, such as sulfur and copper, have been studied in ZnO Qdots. Increased intensity and changes in spectral distribution of the broad green defect emission with S doping has been reported.

Doped ZnS Qdots are very important semiconductor nanomaterials, with Mn^{2+}-doped ZnS Qdots being one of those most studied as a phosphor. In 1994, researchers reported very high PL-QY (~18%) from ZnS:Mn Qdots. Coincident with the intensity enhancement, they reported shorter luminescent lifetimes for the Mn^{2+} emission (decrease from hundreds of microseconds for the bulk to nanoseconds in nanocrystals). The increased intensity was attributed to an efficient energy transfer from the ZnS host to Mn^{2+} ions facilitated by mixed electronic states. Hybridization of atomic orbitals of ZnS and d-orbitals of Mn^{2+} in the nanoparticles was suggested to also be responsible for the relaxation of selection rules for the spin-forbidden $^4T_1 \rightarrow {}^6A_1$ transition of Mn^{2+}, leading to the short emission lifetimes. Subsequent research demonstrated that while the QY of passivated ZnS:Mn Qdots could be high, the luminescent lifetimes were not significantly smaller from those of the bulk material. The luminescence properties were, however, found to be dependent upon the S^{2-} and Mn^{2+} concentrations, as well as, the structural properties of the Qdots. The $^4T_1 \rightarrow {}^6A_1$ Mn^{2+} emission intensity generally increases with increasing doping Mn^{2+} concentration and a quenching of Mn^{2+} emission was observed at high Mn^{2+} concentrations (>0.12 at %). The local environment around the Mn^{2+} in the Qdots has been studied using X-ray absorption fine structure (XAFS) and electron spin resonance (ESR). XAFS data showed that the Mn^{2+} substituted on the tetrahedral Zn^{2+} site in the lattice. ESR data were consistent with this conclusion, showing a spectrum for Mn^{2+} spins typical of a tetrahedral crystal field. In some cases for ZnS:Mn Qdots, the ESR spectra show a Mn^{2+} signal with octahedral symmetry, but the location of this defect site is not fully understood. It has also been suggested that this signal resulted from Mn^{2+} on the surface of the Qdots versus in the interior, but this assignment has been disputed and attributed to Mn-Mn clustering at high concentrations.

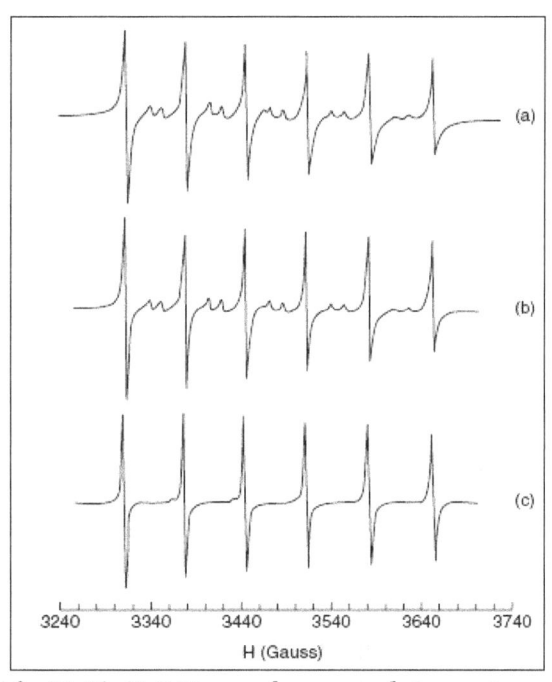

Figure: EPR spectra for Mn^{2+} in ZnS:Mn sample measured at room temperature; (a), 0.003% Mn^{2+}(experimental), (b) 0.008% Mn^{2+}(experimental), and (c) 0.008% Mn^{2+}.

Preparation of Quantum Dots

What is a Colloid?

A colloid is one of the three primary types of mixtures, with the other two being a solution and suspension. A colloid is a solution that has particles ranging between 1 and 1000 nanometers in diameter, yet is still able to remain evenly distributed throughout the solution. These are also known as colloidal dispersions because the substances remain dispersed and do not settle to the bottom of the container. In colloids, one substance is evenly dispersed in another. The substance being dispersed is referred to as being in the dispersed phase, while the substance in which it is dispersed is in the continuous phase.

When a dispersed phase is dispersed in a dispersion medium then depending on the degree of dispersion, the systems are classed as,

- True solution,
- Colloidal solution,
- Suspension.

Properties	True solution	Colloidal solution	Suspension
Particle size	1 Å – 10 Å	10 Å – 1000 Å	More than 1000 Å
Appearance	Clear	Generally clear	Opaque
Nature	Homogeneous	Heterogeneous	Heterogeneous
Separation by filtration	Not possible	Not possible	Possible
Separation by cellophane paper	Not possible	Possible	Possible
Visibility	Not visible under microscope	Visible under ultra-microscope	Visible to naked eye
Brownian motion	Not observable	Occurs	May occur

Example of Colloids

Colloidal AgCl, AgI, Ag proteinate (effective germicide), colloidal sulphur. Many natural and synthetic polymers are important in pharmaceutical practice.

Polymers are the macromolecules formed by polymerization or condensation of small non-colloidal molecules e.g. proteins, natural colloids, plasma proteins which are responsible for binding certain drug molecules so that the pharmacological action of the drug molecule is affected by them. Starch and hydroxymethylallulose, cyclodeztrin are also examples.

Dispersion Medium	Dispersed Phase	Type of Colloid	Example
Solid	Solid	Solid sol	Ruby glass
Solid	Liquid	Solid emulsion/gel	Pearl, cheese
Solid	Gas	Solid foam	Lava, pumice
Liquid	Solid	Sol	Paints, cell fluids
Liquid	Liquid	Emulsion	Milk, oil in water
Liquid	Gas	Foam	Soap suds, whipped cream
Gas	Solid	Aerosol	Smoke
Gas	Liquid	Aerosol	Fog, mist

Types of Colloids

- Lyophilic colloids (solvent loving): They are so called because of affinity of particles for the dispersion medium. Solutions of lyophiles are prepared by simply dissolving the material in the solvent. Because of attraction between the dispersed phase and dispersion medium, salvation (hydration in case of water) of the particles occurs. Most of these colloids are organic n nature e.g. gelatin, acacia, insulin, albumin. The solution is viscous because of strong affinity for water (called gels).

- Lyophobic colloid (solvent hating): The dispersed phase has little attraction to the solvent (solvent hating). Their properties differ from the lyophilic (hydrophilic). They are usually inorganic n nature e.g. gold, silver, sulphur. In contrast to lyophilic colloid, it is necessary to use special method to prepare hydrophobic colloid.

- Hydrophilic sol: For lyophilic sol when the dispersion medium is water then it is called then they are called hydrophilic sols. Such as starch, glue, proteins, gelatin and certain other organic compounds.

- Hydrophobic sols: For lyophobic sol when the dispersion medium is water then it is called then they are called hydrophobic sols. Examples are sol of metals, metal sulphides, metal hydroxides, sulphur, phosphorous and other inorganic substances.

Properties	Lyophobic sols or Hydrophobic sol	Lyophilic sols or hydrophilic sol
Detection of particles	The particles may be readily detected by means of an ultra-microscope.	The particles are not detected by means of an ultra-microscope.
Viscosity	Hardly differs from that of the dispersion medium.	Much higher than that of the dispersion medium.
Electric charge	All particles in a sol have the same charge resulting from the adsorption of ions from solution.	The charge on colloidal particles depends upon the pH of the medium, since the particles readily adsorb H+ or OH- ions. This charge is often due to the dissociation of the molecules of the disperse substance.

Migration of particles in the electric field	The particles migrates in one characteristic direction depending on the charge they bear.	The particles may migrate in either direction or may not migrate at all, depending on the pH of the medium.
Stability	Dispersed particles are precipitated by the addition of small amount of an electrolyte.	Dispersed particles are not precipitated by small amounts of electrolytes although large quantities cause precipitation.
Nature	When the liquid is removed, the resulting solid does not form sol again by the simple addition of the liquid.	When the liquid is removed, resulting jelly-like solid is recoverted into sol by the addition of the liquid.
Occurrence	Generally, do not occur naturally.	Most of these occur naturally.

Interaction between Colloidal Particles

The forces responsible for interactions of colloidal particles can be summarized as:

- Excluded Volume Repulsion: This obscures any overlap between hard particles.

- Electrostatic Interaction: Usually the colloid particles carry electrical charges. This causes attractive/repulsive Coulombic interactions among particles. These interactions are affected by the charges of both the dispersed phase and dispersion medium, and also on the mobility of these phases.

- Van der Waals Forces: These occur because of the interactions between two dipoles, which may be either permanent or induced. Although the colloid particles do not have permanent dipoles, any variations in electron density may cause temporary dipoles in the particles. A temporary dipole in one particle can induce dipole in the nearby particles as well. Both these dipoles (temporary and induced) are attracted towards each other, owing to the van der Waals forces. These forces are always present (unless the dispersed and continuous phases have same refractive index). These forces are short range and are attractive.

- Entropic Forces: Second law of thermodynamics suggests that a system progresses to the state of maximum entropy. This results in some net force between almost all particles.

- Steric Forces: These forces modulate the inter-particle forces between polymer coated surfaces or in solutions having non-adsorbing polymer. They result in either steric repulsive forces (mostly of entropic origin) or attractive depletion forces between them. These effects are of high importance in super plasticizers wherein they can boost the workability of concrete and also reduce water content.

Preparation of Colloids

We have two main types of methods for the preparation of colloidal solutions: Dispersion and Condensation.

Dispersion Method

In the dispersion or disintegration methods, as the name suggests, particles of colloidal size are produced by disintegration of a bulk quantity of a hydrophobic material. These methods may involve the use of such mechanical methods as:

- Mechanical dispersion,
- Electro-dispersion,
- Ultrasonic dispersion,
- Peptization.

Mechanical Dispersion

The substance to be dispersed is ground as finely as possible by the usual methods. It is shaken with the dispersion medium and thus obtained in the form of a coarse suspension. This suspension is now passed through a colloid mill. The simplest type of colloid mill called disc mill, consists of two metal discs nearly touching each other and rotating in opposite directions at a very high speed. The suspension passing through these rotating discs is exposed to a powerful shearing force and the suspended particles are apart to yield particles of colloidal size. Colloid mill are widely used in the industrial preparation of paints, cement, food products, pharmaceutical products etc.

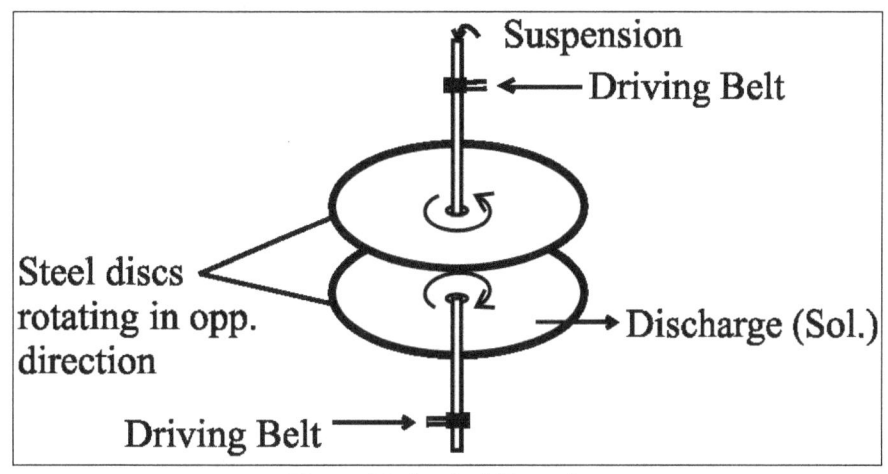

Figure: Mechanical dispersion.

Electro-dispersion

These methods are employed for obtaining colloidal solutions of metals like gold, silver, platinum etc. An electric arc is struck between the two metallic electrodes placed in a container of water. The intense heat of the arc converts the metal into vapors, which are condensed immediately in the cold water bath. This results in the formation of particles of colloidal size. We call it as gold sol.

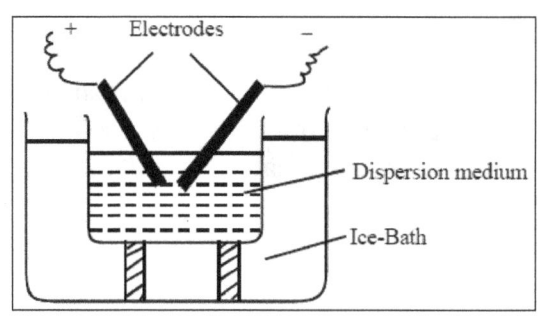

Figure: Preparation of colloidal solution by Bredig's Arc method.

Ultrasonic Dispersion

Ultrasonic vibrations (having frequency more than the frequency of audible sound) could bring about the transformation of coarse suspension to colloidal dimensions. Claus obtained mercury sol by subjecting mercury to sufficiently high frequency ultrasonic vibration.

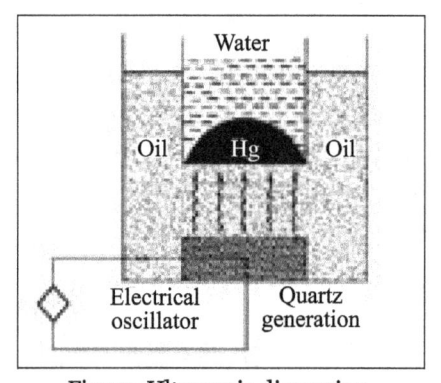

Figure: Ultrasonic dispersion.

Peptization

Peptization is the process of converting a freshly prepared precipitate into colloidal form by the addition of a suitable electrolyte. The electrolyte is called peptizing agent. For example when ferric chloride is added to a precipitate of ferric hydroxide, ferric hydroxide gets converted into reddish brown colored colloidal solution. This is due to preferential adsorption of cations of the electrolyte by the precipitate. When $FeCl_3$ is added to $Fe(OH)_3$, Fe^{3+} ions from $FeCl_3$ are adsorbed by $Fe(OH)_3$ particles. Thus the $Fe(OH)_3$ particles acquire + ve charge and they start repelling each other forming a colloidal solution.

Condensation Methods

Sulphur sol is obtained by bubbling H_2S gas through the solution of an oxidizing agent like HNO_3 or Br_2 water, etc. according to the following equation:

$$Br_2 + H_2S \rightarrow S + 2HBr$$

$$2HNO_3 + H_2S \rightarrow 2H_2O + 2NO_2 + S$$

Fe(OH)$_3$ sol, As$_2$S$_3$ sol can also be prepared by chemical methods.

Purification of Colloids

When a colloidal solution is prepared it contains certain impurities. These impurities are mainly electrolytic in nature and they tend to destabilize the colloidal solutions. Therefore colloidal solutions are purified by the following methods:

- Dialysis: The process of dialysis is based on the fact that colloidal particles cannot pass through parchment or celloplane membrane while the ions of the electrolyte can. The colloidal solution is taken in a bag of cellophane which is suspended in a tub full of fresh water. The impurities diffuse out leaving pure colloidal solution in the bag. This process of separating the particles of colloids from impurities by means of diffusion through a suitable membrane is called dialysis.

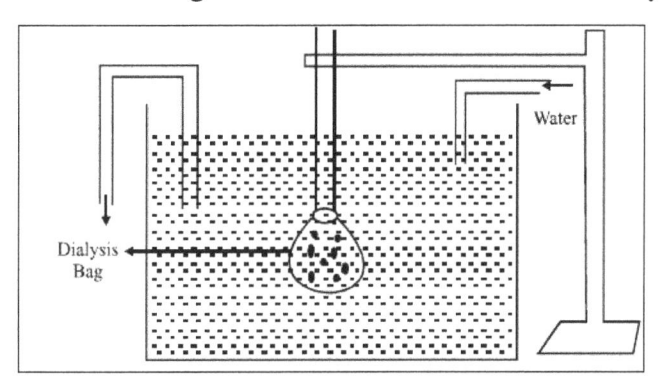

- Electro-dialysis: The dialysis process is slow and to speed up its rate, it is carried out in the presence of an electrical field. When the electric field is applied through the electrodes, the ions of the electrolyte present as impurity diffuse towards oppositely charged electrodes at a fast rate. The dialysis carried out in the presence of electric field is known as electro-dialysis.

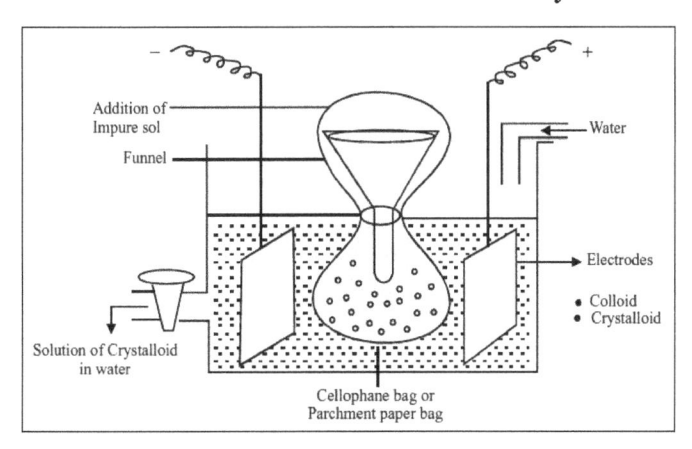

Stability of Colloids

Colloidal particles, though larger than ions and molecules, yet are stable, and do not settle under gravity. There are at least three good reasons for the stability of colloidal sols:

- Brownian motion: Like the molecules or ions in a solution, the colloidal particles of a sol are in a state of continuous rapid motion. The intensity of Brownian motion falls rapidly with increase in the particle size, yet it is high enough to offset of gravity in case of colloidal particles.

- Electric charge: As we know that the colloidal particles in a sol are all either positively charged or negatively charged. Therefore, the force of repulsion keeps the particles scattered and even upon close approach they will not collide and coalesce. Hence similar charge on all the particles of a colloid accounts for the stability due to mutual repulsion in the solution.

- Solvation: The colloidal particles of a sol are often highly hydrated in solution. The resulting hydrated "shell" prevents close contact and cohesion od colloidal particles. Comparatively the addition of small amounts of a lyophilic colloid called protective colloids.

Synthesis of Quantum Dots

Micellar Synthesis of Colloidal Quantum Dots

Micellar synthesis of colloidal quantum dots Water-in-oil, a reverse micro-emulsion technique, to prepare colloidal quantum dots was first proposed in the last two decades of the twentieth century. The technique involves a chemical reaction which can be controlled by inter-micellar exchange of reactants, resulting in nucleation and growth of the nanoparticles. This process has been employed to produce a variety of nanoparticles from different substances, such as metals, metallic oxides, silver halides, and semiconductors. Earlier, nanoparticle growth via reverse micro emulsions was assumed to be limited by reverse micelle shells. As reverse micelle are monodisperse, the nanoparticles, thus produced, should also be monodisperse. Based on this logic, the size of the nanoparticles could be altered since the size of reverse micelles is controllable by variations in the concentrations of water and surfactants.

Hot-solution Decomposition Process

Precursors, such as alkyl, acetate, carbonate and oxides of Group II elements, are mixed with Group VI phosphene or bis(trimethyl-silyl) precursors. A typical procedure involves first degassing and drying of trioctyl-phosphine oxide (TOPO, a coordinating solvent) at 200–350°C under vacuum (1 Torr or 7.5×10^{-6} Pa) in a three-neck round flask in a dry box. A mixture of Cd-precursor and tri-n-octyl-phosphine (TOP) selenide is prepared in a dry box and injected with vigorous stirring into the flask at a temperature of ~300°C. The simultaneous injection of precursors into the flask along with TOPO results in homogeneous nucleation to form Qdots, with the subsequent growth of Qdots through 'Ostwald ripening' being relatively slow. In Ostwald ripening, the higher free energy of smaller Qdots makes them lose mass to large size Qdots, eventually disappearing. The net result is a slow increase of the size of Qdots at the reaction temperature of ~230–250°C (depending on precursor, coordinating agents and solvents). The coordinating TOPO solvent stabilizes the Qdot dispersion, improves the passivation of the surface, and provides an adsorption barrier to slow the growth of the Qdots. The final size of the Qdots is mainly controlled by the reaction time and temperature. Aliquots may be removed from the flask at regular intervals during the first few hours and the optical absorption edge used to achieve a desired particle size. This method has been extensively used to synthesize II-VI, IV-VI and III-V Qdots. The size, shape, and control of the overall reaction depends not only on process parameters and precursors, solvents and coordinating agents, but also on the purity of the coordinating solvent, such as TOPO. It has been reported that technical grade TOPO (90% pure) were better for synthesizing uniform Qdots than the pure TOPO.

An advantage of this synthesis route is that it provides sufficient thermal energy to anneal defects and results in monodispersed Qdots (typically standard deviation about the average size of 5%). Since growth of the particles in this process is relatively slow and can be controlled by modulating the temperature, a series of Qdot sizes can be prepared from the same precursor bath. Using this process large quantities of Qdots and alloying process have been demonstrated. Some of the disadvantages of this method include higher costs due to the use of high temperature, toxicity of some of the organometallic precursors, and generally poor dispersions in water. Table below shows a chronological summary of this synthesis technique with different growth parameters and precursors to produce different Qdots.

Table: Synthesis of different sized Qdots using hot-solution decomposition reaction.

Qdots	Precursor	Process parameters	Particle Size (nm)
GaAs	$GaCl_3$, $(TMS)_3As$ in Quinoline	240 °C for 3days; flame anneal at 450 °C.	2.4
ZnS, ZnSe, CdS, CdSe, CdTe, HgTe	$M(ER)_2$; R: n-butyl phenyl; E: S, Se, Te; M: Cd, Zn, Hg and/or phosphine complexes; Co.Sol.: DEPE	DEPE and $M(ER)_2$ reacted, (Temp. range: 250–300 °C).	2.5–5 nm
CdS, CdSe, CdTe	Me_2Cd, silylchalconides, Phosphine chalconides; Co.sol: TOPO & TOP/TBP	300–350 °C at 1 atm at Ar (TOPO degassing); 230–260 °C (growth temp.).	1.2–11.5
GaAs	$GaCl_3$/ GaI_3, diglyme, As, toluene, Na-K alloy	As, Na-K alloy mixture refluxed to 100 °C in Ar for 2 days; $GaCl_3$/GaI_3 diglyme mixture added, heated from 0 °C to RT to 111 °C. for 2 days.	6–10

InP, GaP, GaInP$_2$	Mixture of chloroindium/gallium oxalate (GaCl$_3$ for GaP) and (TMS)$_3$P in CH$_3$CN; Co.sol: TOPO & TOP	270–360°C at airless condition for 3 days; Qdots dispersed in methanol.	2.6–4.6 (InP), 3 (GaP), 6.5 (GaInP$_2$)
InP, InAs	InCl3, TOPO, (TMS)$_3$P/(TMS)$_3$As	InCl3 & TOPO heat at 100°C for 12 h, (TMS)$_3$P added, after 3hr heated to 265°C for 6 days.	2–6
CdSe/ZnS	Me$_2$Cd, Me$_2$Zn, Se, (TMS)2S, Co.sol: TOPO, TOP	Single step synthesis Core: 350 °C at 1 atm at Ar, growth: 310 °C Shell: 300 °C.	2.7–4
CdSe/ZnS	Me$_2$Cd, Me$_2$Zn, Se, (TMS)$_2$S, Co.sol: TOPO, TOP	Two step synthesis (airless) Core growth: 290–300°C Shell growth: 140°C for 2.3 nm & 220°C for 5.5 nm.	2.3–5.5
CdSe/CdS	Me$_2$Cd, Se, (TMS)$_2$S, Co.sol: TOPO, TBP	Two step process: Core: 300°C; Shell: 100°C.	2.5–4
ZnSe	Me$_2$Zn, Se, HDA, TOP	HDA dried & degassed at 150°C for hrs in vacuum and heated to 310 °C at 1 atm in Ar; Core growth with Zn & Se precursor at 270°C.	4.3–6 nm
InAs/InP InAs/CdSe	(TMS)3As, Indium (III) chloride, TOP (TMS)3P, Me$_2$Cd; TBPSe	Two-step Process (airless) Core growth: 260°C; Shell: dropwise addition; 260°C.	2.5–6 nm (InAs); 1.7 (core/shell)
CdSe	Me$_2$Cd, Se, TBP, TOPO, HPA	TOPO (+HPA 1.5–3 wt%) degassed at 360°C (or 310 °C, 280 °C); Core growth: 300 °C (or 280 °C or 250°C).	~6nm
ZnSe:Mn	Me$_2$Mn, Et$_2$Zn, TOP, Se, HDA	Dimethyl Mn, TOP, Se, Diethyl Zn mixture added to HDA at 10°C in N$_2$. Growth: 240–300°C.	2.7–6.3
CdSe/ZnS	Me$_2$Cd, Se, TOP, TOPO, HDA, (TMS)$_2$S, Me$_2$Zn	Two Step: Core: reaction & growth: 270–310 °C; Shell: slow addition of Zn & S precursor at 180–220 °C.	4.5–5 nm
CdSe	Scheme 1: Cd(Ac)$_2$, SA/TOPO; 2: Cd(Ac)$_2$, SA; 3: CdCO$_3$, SA/TOPO; 4: CdCO$_3$, LA/TOPO; 5: CdO, SA/TOPO; 6: Cd(Ac)$_2$, tech TOPO; 7: CdO, TDPA/TOPO	Solvent & Cd-precursor heated to 250– 360° C at Ar; TOP-Se or TBP-Se injected; Growth temp: 200–320 °C (if DDA involve, temp: ~220 °C.	2–25nm
CdS, CdSe, CdTe	CdO, TOPO, HPA/TDPA, S, Se, Te & TOP	One pot: CdO, HPA/TDPA heated 300 °C; Core with chalconide precursor: reaction: 270 °C, & growth 250 °C.	2-8nm

CdSe	CdO, Se, TOPO, TBP, HDA, ODA, SA	CdO & SA, heated to 150 °C in Ar; after CdO dissolution, cool to RT; TOPO & HAD added & heated to 320 °C in Ar; TBP-Se added, Growth 290 °C.	
PbS	PbO, OA, (TMS)2S, TOP	PbO dissolved in oleic acid at 150 °C in Ar; $(TMS)_2S$ & TOP injected.	5nm
CdSeS	CdO, OA, TOA, Se, S, TOP	CdO+ OA+TOA heated at 300 °C in N_2, TOP-S, TOP-Se injected.	~5nm
PbSe	Pb-acetate trihydrate, OA, Se, TOP	Single Step: Pb acetate + Co.sol degassed at 100–120 °C at 300–500 mTorr for 2h; reaction and growth: 140 °C.	5 nm
CdSe	CdO, OA, TOA, C8SH or C18SH,	CdO + OA + TOA heated at 300 °C; TOA + C8SH or C18SH injected.	3, 4, 6 nm

Ac: acetate; Co.sol: coordinating solvent; DDA: dodecylamine; DEPE: 1,2-bis(diethyl-phosphino)-ethane; DMPA: 2,2 –dimethoxy-2-phenylacetophenone; EGDMA: ethylene glycol domethacrylate; Et: ethyl; HDA: hexadecylamine; HPA: hexyl-phosphonic acid; LA: lauric acid; Me: methyl; MMA: methylmethacrylate; MPA: marcaptopropionic acid; OA: oleic acid; ODA: octadecylamine; ODE: 1-octadecene; SA: stearic acid; TBP: tri-n-butyl phosphine; TDPA: tetradecylphosphonic acid; TMS: trimethyl-silyl; TOA: trioctyl amine; TOP: tri-n-octyl-phosphine; TOPO: tri-n-octyl-phophine oxide (tech TOPO: Technical grade TOPO) 1 Torr = 7.5×10^{-6} Pa.

Sol-gel Process

Sol-gel techniques have been used for many years to synthesize nanoparticles including Qdots. In a typical technique, a sol (nanoparticles dispersed in a solvent by Brownian motion) is prepared using a metal precursor (generally alkoxides, acetates or nitrates) in an acidic or basic medium. The three main steps in this process are hydrolysis, condensation (sol formation) and growth (gel formation). In brief, the metal precursor hydrolyzes in the medium and condenses to form a sol, followed by polymerization to form a network (gel). This method has been used to synthesize II-VI & IV-VI Qdots, such as CdS, ZnO, PbS. As an example, ZnO Qdots have been prepared by mixing solutions of Zn-acetate in alcohol and sodium hydroxide, followed by control aging in air. The process is simple, cost-effective and suitable for scale-up. The main disadvantages of the sol-gel process include a broad size distribution and a high concentration of defects. Therefore, this synthesis technique is used sparingly.

Vapor-phase Methods

These vapor-phase methods for producing Qdots begin with processes in which layers are grown in an atom-by-atom process. Consequently, self-assembly of Qdots occurs on a Materials 2010, 3 2297 substrate without any patterning. In general, the layered materials grow as a uniform, often epitaxial layer (Frank-van der Merwe mode–FvdM), initially as a smooth layer sometimes followed by nucleation and growth of small islands (Volmer-Weber mode–VW), or as small (Qdots) islands directly on the substrate (Stranski-Krastonow mode–SK). Depending on the interfacial/surface energies and lattice mismatch (i.e., lattice strain), one of these growth modes is observed. For example, Qdots may be formed by SK growth with an over layer material that has a good lattice match with the substrate, but the substrate surface energy (σ_1) is less than the sum of the interfacial energy between the substrate and over layer (γ_{12}) and the over layer surface energy (σ_2), i.e., when $\sigma_1 < \sigma_2 + \gamma_{12}$. In other cases, formation of Qdots was due to relaxation of strain required to maintain epitaxy. In the case of substrates with an over layer with a large lattice mismatch and appropriately small surface and interface energies, initial growth of the over layer occurs through by a layer-by-layer FvdM growth. However, when the film is sufficiently thick (a few monolayers) to induce a large strain energy, the system lowers its total free energy by breaking the film into isolated islands or Qdots (i.e., the VW mode). Kim et al. synthesized ZnSe/ZnS Qdots with the SK growth mode using a metal-organic chemical vapor deposition (CVD) technique in an atomic layer epitaxy (ALE) mode. The mean dot height was 1–1.9 nm. An apparent temperature dependent, anomalous behavior of confined carriers in the ZnSe Qdots was observed and attributed to thermalized carrier hopping between Qdots. The carrier hopping resulted in a substantial decrease of the PL peak energy and line width when the sample was cooled from room temperature to 140 K.

Molecular beam epitaxy (MBE) has been used to deposit the over layers and grow elemental, compound or alloy semiconductor nanostructured materials on a heated substrate under ultra-high vacuum ($\sim 10^{-10}$ Torr or 7.5×10^{-16} Pa) conditions. The basic principle of the MBE process is evaporation from an apertured source (Knudsen effusion cell) to form a beam of atoms or molecules. The beams in the MBE process can be formed from solids (e.g., elemental Ga and As are used to produce GaAs Qdots) or a combination of solid plus gases (e.g., AsH_3, PH_3, or metal-organics such as tri-methyl gallium or tri-ethyl gallium). The metal-organic sources may leave high concentrations of carbon in the Qdots. MBE has been mainly used to self-assemble of Qdots from III-V semiconductors and II-VI semiconductors using the large lattice mismatch e.g., InAs on GaAs has a 7% mismatch and leads to SK growth.

Layer growth by physical vapor deposition (PVD) results from condensation of a solid from vapors produced by thermal evaporation or by sputtering. Different techniques have been used to cause evaporation, such as electron beam heating, resistive or Joule heating, arc-discharge and pulsed laser ablation. CVD is another method to form thin films from which Qdots can be self-assembled. In CVD, precursors are introduced in a chamber at a particular pressure and temperature and they diffuse to the heated substrate, react to form a film, followed by gas-phase byproducts desorbing from the substrate and being removed from the chamber. InGaAs and AlInAs Qdots have been synthesized using either surface energy or strained-induced SK growth processes. Although, self-assembling of Qdots using vapor-phase methods is effective in producing Qdots arrays without template, fluctuation in size of Qdots often results in inhomogeneous optoelectronic properties.

Applications of Quantum Dots

Solar Cells

One of the most important features of silicon of use in electronic devices (e.g. transistors) is doping. Wherein small quantities of various elements can be added to silicon in order to generate either an excess (n-type) or deficiency (p-type) of electrons in it, thereby enhancing the material's conductivity. Additionally, p-n junction can be created by joining n-type and p-type silicon, which are the building blocks for almost all the devices in electronic industry.

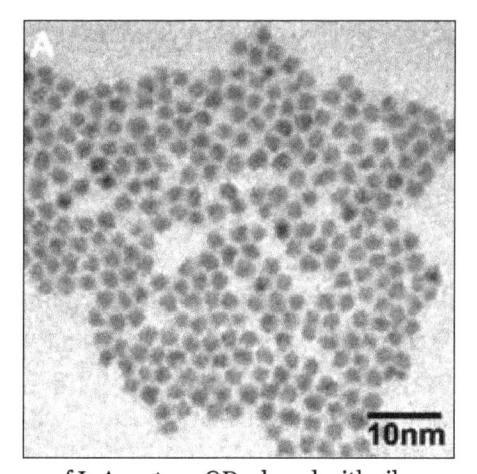

Figure: TEM images of InAs n-type QDs doped with silver, 3.3nm in diameter.

Doping the QDs is a relatively new area of research, and both n-type and p-type QDs have been produced. Figure above shows the indium arsenide QDs which have been doped with silver to create n-type QD. Low cost solar cells can be realized by using QDs, as QDs can be easily prepared via simple and economic chemical processes. Additionally, thin-film photo voltaics can be prepared from QDs which have efficiencies comparable to those of traditional silicon cells. The enhancements in efficiency of QDs can be attributed to the different materials that can be used in their synthesis. Since some semiconductors can emit multiple electrons on absorbing one photon. In addition to this, the manipulation in their size and shape can result in absorption of light having different colors.

However, the synthesis of efficient QD based solar cells has not yet been possible. For preparing solar cell, an n-type, and a p-type nanocrystals are required. When light is incident on a solar cell, electrons and holes are generated as light photons get absorbed in the material. These electrons and holes must be separated so asto avoid their spontaneous recombination. The separation of these charge carriers results in the flow of electrons out of the semiconductor to the external electrical circuit. However, some of the electrons and holes recombine, this does not cause and electron (and thereby current) flow in the external circuit. This recombination is much pronounced in QDs than the large silicon crystals. Doping of the semiconductor nanocrystals for making p-n junction can separate the electrons and holes more efficiently. Conventionally, silicon is doped with P or B atoms; however, this cannot be done with the QDs because of their nanometer size. For comparison, a 4 nm QD comprises ~1000 atoms. Addition of few dopant atoms can result in expelling these atoms from the nanocrystals.

QLEDs

Qdots-based light emitting diodes (QLEDs) have attracted intense research and commercialization efforts over the last decade. Infact, QLEDs have several advantages compared to organic LEDs (OLEDs). These are as follows:

- FWHM of the emission peak from Qdots is only 20–30 nm, compared with >50 nm for their organic counterpart, which is necessary for a high quality image.

- Inorganic materials usually show better thermal stability than organic materials. Under operating at high brightness as well and/or high current, Joule heat is one of the predominant problems for device degradation. With better thermal stability, inorganic materials based devices are expected to exhibit longer lifetimes.

- The display color of OLEDs generally changes with time due to the different lifetime of the red, green and blue pixels. However, one can obtain all of three premium colors from Qdots with the same composition changing the particle size (due to the quantum confinement effect, as discussed above). The same chemical composition should exhibit similar degradation with time.

- The QLED device can produce IR emission while the organic molecules in OLEDs usually exhibit wavelengths shorter than 1 μm.

- The spin statistics are not restrictive for Qdots, i.e., external quantum efficiency (EQE) of 100% can be achieved (in principle).

The EQE of QLEDs can be expressed as: $\eta_{Ext} = \eta_r \cdot \eta_{INT} \cdot \eta \cdot \eta_{OUT}$, where, ηr is the probability of holes and electron forming exciton, η_{INT} is the internal PL-QY and η & η_{OUT} are the probability of radiative decay and out-coupling efficiency, respectively. The value of η_r for fluorescent organics theoretically is limited to 25% as the ratio of singlet to triplet states is 1:3 and only singlet state recombinations result in luminescence. However, for phosphorescent organics it is found > 25% due to spin-orbit coupling. Note that phosphorescent organics lead to rapid degradation of the host material. With respect to high η_{Ext}, the value of η_{OUT} for planar devices is typically found to be ~20%. The η_{OUT} efficiency can be enhanced by incorporating a microcavity structure. For QLEDs, the value of η_{INT} (QY) can approach 100% and for a device with the appropriate electron and hole energies, the value of η_r can also be ~100%. It has been observed that a QLED can emit light under both forward and reverse bias. Reason for this behavior is uncertain and many explanation are plausible, such as, different rates of electron and hole injections, different carrier mobility in the electron and hole transport layers, energy level offsets of the different layers, the emitting layer composition, surface and, uniformity and thickness of the layers. The valence (HOMO) and conduction (LUMO) band energies of some of the polymers that have been used in inorganic/organic QLEDs and solar cell are tabulated in Table below. Another table shows the valence band and conduction band-energies reported in literature for different sized and structured Qdots, and Table below shows the work functions of some of the commonly used materials in QLED solar cell and OLEDs.

Table: Valence band (HOMO) and conduction band (LUMO) energies for some of the commonly used organics for QLEDs, solar cells and OLEDs.

Organics	Conduction Band (eV)	Valence Band (eV)
Alq3	3.1	5.8
CBP	2.9	6.0
PBD	2.6	6.1
PCBM	4.0	6.5
PPV	2.5	5.1
PVK	2.2	5.3
TAZ	3.0	6.5
TFB	2.2	5.4
TPBI	2.7	6.2
TPD	2.1	5.4
Poly TPD	2.3/2.5	5.2/5.4

Alq3: tris-(8-hydroxyquinoline) aluminum; CBP: 4,4',N,N'-diphenylcarbazole; t-Bu-PBD: 2-(4-biphenylyl)- 5-(4-tert-butylphenyl)-1,3,4 oxadiazole; PCBM: [6,6]-phenyl C61 butyric acid methyl ester; PPV: poly(phenylene vinylene); PVK: poly(vinyl-carbazole); TAZ: 3-(4-Biphenylyl)- 4-phenyl-5-tert-butylphenyl1, 2, 4-triazole; TFB: Poly[(9,9-dioctylfluorenyl-2,7odiyl)-co-(4-4'-(N-(4-sec-butylphenyl)) diphenylamine)]; TPBI: 1,3,5-tris(N-phenylbenzimidazole-2-yl)-benzene; TPD: N, N'-diphenyl-N, N'-bis(3-methylphenyl)-(1, 1'-biphenyl)-4, 4'-diamine.

Table: Valence band and conduction band energies for different sized and structured CdSe Qdots.

Qdots	Conduction Band (eV)	Valence Band (eV)	Particle Size (nm)	Emission
CdSe	4.4	6.5	5	
CdSe/CdS	4.4	6.5	4.6	
CdSe/CdS	4.7	6.8	4	
CdSe/CdS/ZnS	4.8	6.8	6.8	600
CdSe/ZnS	4.4	6.5		
CdSe/ZnS	4.3	6.5		550nm
CdSe/ZnS	4.8	6.5		
CdSe/ZnS	4.6 (CdSe)	6.8 (CdSe)	5.8	
CdSe/ZnS	4.7	6.7		
CdSe/ZnS/CdS	3.9	6.0	3-8.3	G,Y,O,R

Table: Work function of some of the commonly used materials as anode or cathode in QLED, solar cell and OLEDs.

Materials	Work functions (eV)
Al	4.1
Ag	4.6
Ca	2.9
ITO	4.7
LiF/Al	2.8
Mg	3.7
PEDOT:PSS	5

ITO: indium tin oxide; PEDOT:PSS: poly(3,4-ethylene-dioxy-thiophene) poly(styrene-sulfonate).

Down Conversion of Blue or Ultraviolet Light

Qdots are being used in down conversion of high energy to lower energy light. The advantages of Qdots phosphors over conventional inorganic phosphor and/or organic dyes include the (i) high QY of Qdots, (ii) emission range from UV to visible to NIR, (iii) better stability compared to organics, (iv) narrow FWHM ~30 nm (higher color saturation compared to conventional phosphor with typical FWHM: 50–100 nm), and (v) a large absorption window (band-gap to UV, enabling simultaneous excitation of different size Qdots). The main disadvantage of Qdots is self-absorption resulting from overlap of the absorption spectra from larger Qdots with the emission spectra and emission spectra from smaller Qdots. Down conversion of 425 nm blue (from GaN based commercial LED) and UV light (from Hg lamp) using 2.0, 2.6, 4.6 and 5.6 nm CdSe/ZnS Qdots dispersed in a poly-aurylmethacrylate polymer matrix. The composite was prepared at 70–75 °C for 2 hrs. by mixing as-synthesized TOP capped CdSe/ZnS Qdots and monomer of 1-aurylmethacrylate, then adding ethyleneglycol dimethacrylate and a radical initiator azo-bis-isobutyronitrile for crosslinking. Recently, organics-capped ZnSe Qdots were excited with a near-UV InGaN LED to produce white light with Commission Internationale de l'Eclairage (CIE) x, y chromaticity coordinates of 0.38, 0.41. Down conversion of blue light (455 nm) from an InGaN LED resulted in white light emission with a color rendering index (CRI) of > 90 and CIE of 0.33, 0.33, when a mixture of green and red emitting CdSe/ZnSe Qdots was used, as shown in Figure below. We also summarized some of the literature reports on down conversion of Qdots in Table below.

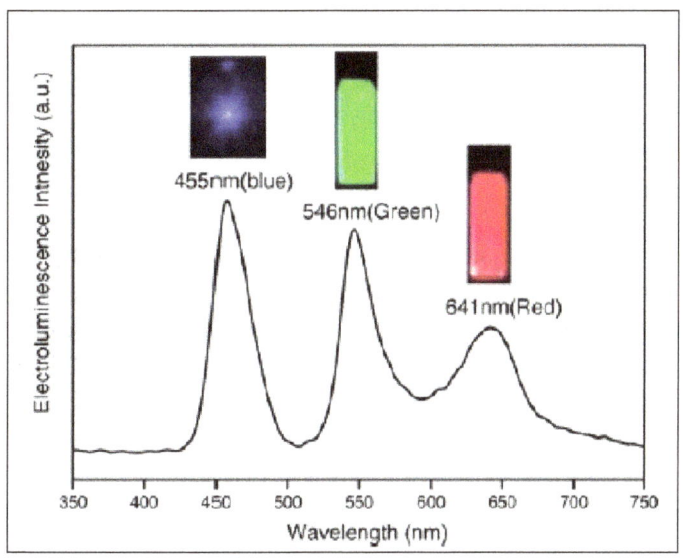

Figure: Down conversion of 455 nm blue emission from an InGaN light emitting diode to green and red by size-tuned CdSe/ZnSe Qdots in a silicone matrix (color rendering index: 91; chormacity coordinate: 0.33, 0.33)

Table: Selected reports on the use of Qdots to down convert blue or UV light from inorganic LEDs.

Source light	Qdots	Matrix	Emitted light
UV (Hg lamp), Blue GaN Commercial LED	CdSe/ZnS (2.0, 2.6, 4.6, 5.6 nm)	Polyaurylmethacrylate	UV: Blue, orange, red; Blue: red (590 nm).
InGaN (near UV)	ZnSe (TOPO & Stearic acid coated)	organics coated ZnSe (10 wt%) dispersed in epoxy resin	White; CIE (0.38, 0.41) Conversion efficiency: 30% relative to RGB commercial phosphors.
InGaN (455 nm	CdSe/ZnSe (G); CdSe/ZnSe (R) CdSe/ZnSe (Y)	TOPO-coated CdSe/ZnSe dispersed in silicone	White, CIE: (0.33, 0.33), CRI: 91 with R& G; White, CIE: (0.32, 0.33), CRI: 50 with Y; Efficiency: 15-30 lm/W.

390 nm UV LED	CdSe/CdS/ZnS	2wt% Qdot in chloroform & epoxy resin at 1:1 (vol); Thermally cured	Red (620 nm).
InGaN/GaN (440 nm, 452 nm)	CdSe/ZnS (440-452 nm)	Qdots blended with resin; 400 -1700 μm (Qdot density: 3.04-140 nanomoles/1ml resin)	White; with 453 nm & CdSe/ZnS (540, 500, 580 & 520 nm): CIE (0.24, 0.33), CRI: 71.
InGaN/GaN (blue/green)	CdSe/ZnS (620nm, R) & Au particles (for surface Plasmon enhanced emission)	5 wt% Qdots and 0.05 wt% Au in toluene spin-coated on LED (thickness ~200 nm)	White: (0.27, 0.24); Conversion efficiency ~53%.

R: red, G: green, B; blue, O: orange; Y: yellow; W; white; CIE: International Commission on illumination; CRI: color rendering index; vol.: by volume.

Nanoelectronics

Nanoelectronics involves using nanotechnology in electronic components. The devices produced are so small that interatomic interactions and quantum confinement effects are applied to them. The devices include hybrid molecular/semiconductor electronics, nanotubes/nanowires (e.g. SiNWs, CNTs), etc. Recent silicon CMOS devices are also within this length scale. Nano electronics is often termed as disruptive technology as present candidates are significantly different from traditional transistors.

Memory Storage

Traditional memory devices use transistors for storing information. With the help of cross bar switches based electronics, ultra high density storage devices can be produced which have reconfigurable interconnections between vertical and horizontal wiring arrays. This has been achieved by Nantero (developed CNTs based crossbar memory 'Nano-RAM'), and Hewlett-Packard (proposed memristor to replace flash memory).

Novel Optoelectronic Devices

Optical and optoelectronic devices offer extremely large bandwidths and high capacities; and are replacing the conventional analog devices in modern communications systems. These include photonics crystals and QDs.

In photonic crystals, the refractive index varies periodically with a lattice constant. The lattice constant is the half of the wavelength of light used. They behave like a semiconductor, with the exception that they work with light or photons, rather than the electrons. Thus, photonic crystals have a tunable band gap for propagating a specific wavelength of light.

QDs are Nano sized objects which can also be used to construct lasers. A QD based laser offers the advantage of tunable emission wavelength over the conventional semiconductor lasers. The emission wavelength can be manipulated by changing the diameter of the QD. Additionally, QD lasers are inexpensive and provide high beam quality.

Quantum Computers

Quantum mechanical principles can be exploited in quantum computers, thereby enabling the use of fast quantum algorithms. Quantum computers use quantum bit (qubit) as memory space for simultaneous multiple computations. Thus much faster computers can be build.

References

- Quantum-Dots-and-Their-Multimodal-Applications-A-Review-229025098: researchgate.net, Retrieved 25, June 2020
- Excitons-and-excitonic-Bohr-radius-energy-levels-splitting: in libnet.ac.in, Retrieved 19, March 2020

4

Understanding Nanostructure

Any structure with at least one dimension measuring in nanometer range is known as nanostructure. This chapter discusses various nanostructures such as nanowire, nanomesh, nanohole, nanosheet, nanopillar, nanostructured film, sculptured thin film, gradient multilayer nanofilm, etc. All the diverse principles of nanostructures have been carefully analyzed in this chapter.

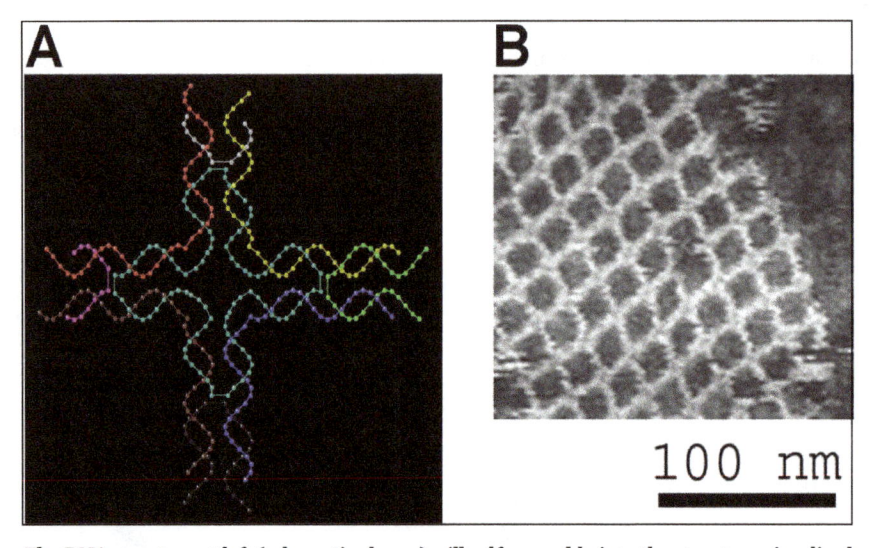

The DNA structure at left (schematic shown) will self-assemble into the structure visualized by atomic force microscopy at right. Image from Strong.

A nanostructure is a structure of intermediate size between microscopic and molecular structures. Nanostructural detail is microstructure at nanoscale.

In describing nanostructures, it is necessary to differentiate between the number of dimensions in the volume of an object which are on the nanoscale. Nanotextured surfaces have *one dimension* on the nanoscale, i.e., only the thickness of the surface of an object is between 0.1 and 100 nm. Nanotubes have *two dimensions* on the nanoscale, i.e., the diameter of the tube is between 0.1 and 100 nm; its length can be far more. Finally, spherical nanoparticles have *three dimensions* on the nanoscale, i.e., the particle is between 0.1 and 100 nm in each spatial dimension. The terms nanoparticles and ultrafine particles (UFP) are often used synonymously although UFP can reach into the micrometre range. The term *nanostructure* is often used when referring to magnetic technology.

Nanoscale structure in biology is often called ultrastructure.

Properties of nanoscale objects and ensembles of these objects are widely studied in physics.

Nanowire

A nanowire is a nanostructure, with the diameter of the order of a nanometer (10^{-9} meters). It can also be defined as the ratio of the length to width being greater than 1000. Alternatively, nanowires can be defined as structures that have a thickness or diameter constrained to tens of nanometers or less and an unconstrained length. At these scales, quantum mechanical effects are important — which coined the term "quantum wires". Many different types of nanowires exist, including superconducting (e.g. YBCO), metallic (e.g. Ni, Pt, Au), semiconducting (e.g. silicon nanowires (SiNWs), InP, GaN) and insulating (e.g. SiO_2, TiO_2). Molecular nanowires are composed of repeating molecular units either organic (e.g. DNA) or inorganic (e.g. $Mo_6S_{9-x}I_x$).

Overview

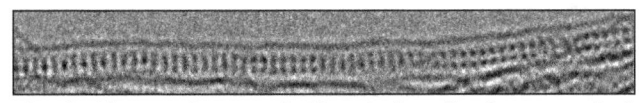

Crystalline 2×2-atom tin selenide nanowire grown inside a single-wall carbon nanotube (tube diameter ~1 nm).

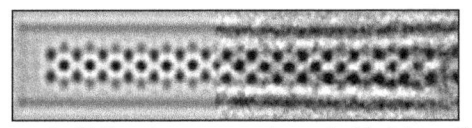

A noise-filtered HRTEM image of a HgTe extreme nanowire embedded down the central pore of a SWCNT. The image is also accompanied by a simulation of the crystal structure.

Typical nanowires exhibit aspect ratios (length-to-width ratio) of 1000 or more. As such they are often referred to as one-dimensional (1-D) materials. Nanowires have many interesting properties that are not seen in bulk or 3-D (three-dimensional) materials. This is because electrons in nanowires are quantum confined laterally and thus occupy energy levels that are different from the traditional continuum of energy levels or bands found in bulk materials.

Peculiar features of this quantum confinement exhibited by certain nanowires manifest themselves in discrete values of the electrical conductance. Such discrete values arise from a quantum mechanical restraint on the number of electrons that can travel through the wire at the nanometer scale. These discrete values are often referred to as the quantum of conductance and are integer multiples of:

$$\frac{2e^2}{h} \simeq 77.41 \, \mu S$$

They are inverse of the well-known resistance unit h/e^2, which is roughly equal to 25812.8 ohms, and referred to as the von Klitzing constant R_K (after Klaus von Klitzing, the discoverer of exact quantization). Since 1990, a fixed conventional value R_{K-90} is accepted.

Examples of nanowires include inorganic molecular nanowires ($Mo_6S_{9-x}I_x$, $Li_2Mo_6Se_6$), which can have a diameter of 0.9 nm and be hundreds of micrometers long. Other important examples are based on semiconductors such as InP, Si, GaN, etc., dielectrics (e.g. SiO_2, TiO_2), or metals (e.g. Ni, Pt).

There are many applications where nanowires may become important in electronic, opto-electronic and nanoelectromechanical devices, as additives in advanced composites, for metallic interconnects in nanoscale quantum devices, as field-emitters and as leads for biomolecular nanosensors.

Synthesis of Nanowires

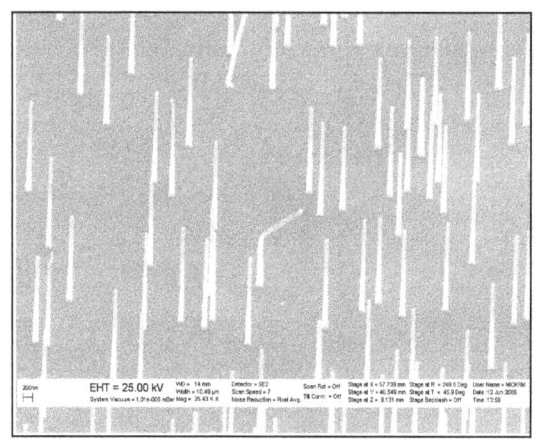

An SEM image of epitaxial nanowire heterostructures grown from catalytic gold nanoparticles.

There are two basic approaches to synthesizing nanowires: top-down and bottom-up. A top-down approach reduces a large piece of material to small pieces, by various means such as lithography or electrophoresis. A bottom-up approach synthesizes the nanowire by combining constituent adatoms. Most synthesis techniques use a bottom-up approach. Initial synthesis via either method may often be followed by a nanowire thermal treatment step, often involving a form of self-limiting oxidation, to fine tune the size and aspect ratio of the structures.

Nanowire production uses several common laboratory techniques, including suspension, electrochemical deposition, vapor deposition, and VLS growth. Ion track technology enables growing homogeneous and segmented nanowires down to 8 nm diameter.

Suspension

A suspended nanowire is a wire produced in a high-vacuum chamber held at the longitudinal extremities. Suspended nanowires can be produced by:

- The chemical etching of a larger wire.
- The bombardment of a larger wire, typically with highly energetic ions.
- Indenting the tip of a STM in the surface of a metal near its melting point, and then retracting it.

VLS Growth

A common technique for creating a nanowire is vapor-liquid-solid method (VLS). This process can produce high-quality crystalline nanowires of many semiconductor materials, for example, VLS–grown single crystalline silicon nanowires (SiNWs) with smooth surfaces could have excellent properties, such as ultra-large elasticity. This method uses a source material from either laser ablated particles or a feed gas such as silane.

VLS synthesis requires a catalyst. For nanowires, the best catalysts are liquid metal (such as gold) nanoclusters, which can either be self-assembled from a thin film by dewetting, or purchased in colloidal form and deposited on a substrate.

The source enters these nanoclusters and begins to saturate them. On reaching supersaturation, the source solidifies and grows outward from the nanocluster. Simply turning off the source can adjust the final length of the nanowire. Switching sources while still in the growth phase can create compound nanowires with super-lattices of alternating materials.

A single-step vapour phase reaction at elevated temperature synthesises inorganic nanowires such as $Mo_6S_{9-x}I_x$. From another point of view, such nanowires are cluster polymers.

Solution-phase Synthesis

Solution-phase synthesis refers to techniques that grow nanowires in solution. They can produce nanowires of many types of materials. Solution-phase synthesis has the advantage that it can produce very large quantities, compared to other methods. In one technique, the polyol synthesis, ethylene glycol is both solvent and reducing agent. This technique is particularly versatile at producing nanowires of gold, lead, platinum, and silver.

The supercritical fluid-liquid-solid growth method can be used to synthesize semiconductor nanowires, e.g., Si and Ge. By using metal nanocrystals as seeds, Si and Ge organometallic precursors are fed into a reactor filled with a supercritical organic solvent, such as toluene. Thermolysis results in degradation of the precursor, allowing release of Si or Ge, and dissolution into the metal nanocrystals. As more of the semiconductor solute is added from the supercritical phase (due to a concentration gradient), a solid crystallite precipitates, and a nanowire grows uniaxially from the nanocrystal seed.

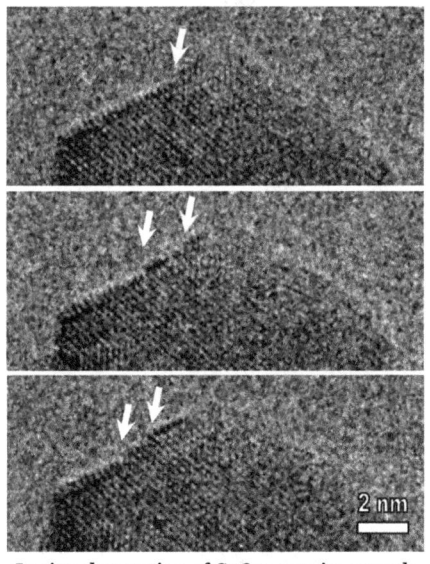

In situ observation of CuO nanowire growth.

Non-catalytic Growth

Nanowires can be also grown without the help of catalysts, which gives an advantage of pure nanowires and minimizes the number of technological steps. The simplest methods to obtain metal oxide nanowires use ordinary heating of the metals, e.g. metal wire heated with battery, by Joule heating in air can be easily done at home. The vast majority of nanowire-formation mechanisms are explained through the use of catalytic nanoparticles, which drive the nanowire growth and are either added intentionally or generated during the growth. However the mechanisms for catalyst-free growth of nanowires (or whiskers) were known from 1950s. Spontaneous nanowire formation by non-catalytic methods were explained by the dislocation present in specific directions or the growth anisotropy of various crystal faces. More recently, after microscopy advancement, the nanowire growth driven by screw dislocations or twin boundaries were demonstrated. The picture on the right shows a single atomic layer growth on the tip of CuO nanowire, observed by in situ TEM microscopy during the non-catalytic synthesis of nanowire.

Physics of Nanowires

Conductivity of Nanowires

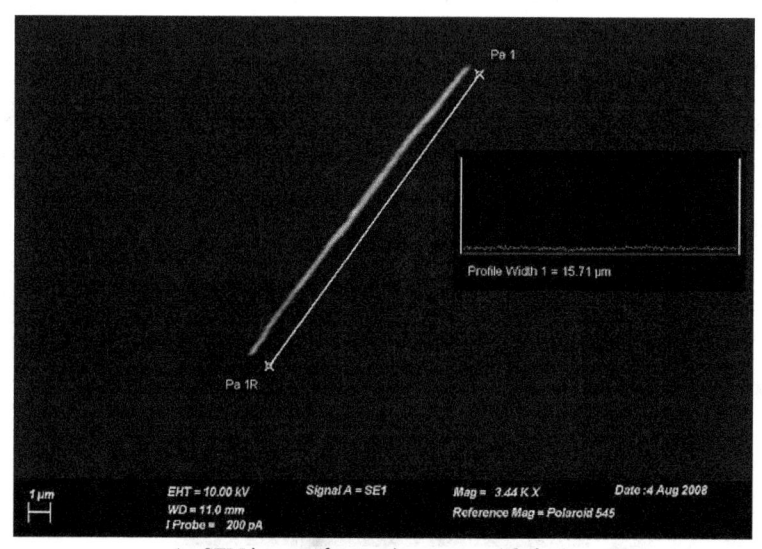

An SEM image of a 15 micrometer nickel wire.

Several physical reasons predict that the conductivity of a nanowire will be much less than that of the corresponding bulk material. First, there is scattering from the wire boundaries, whose effect will be very significant whenever the wire width is below the free electron mean free path of the bulk material. In copper, for example, the mean free path is 40 nm. Copper nanowires less than 40 nm wide will shorten the mean free path to the wire width.

Nanowires also show other peculiar electrical properties due to their size. Unlike single wall carbon nanotubes, whose motion of electrons can fall under the regime of ballistic transport (meaning the electrons can travel freely from one electrode to the other), nanowire conductivity is strongly influenced by edge effects. The edge effects come from atoms that lay at the nanowire surface and are not fully bonded to neighboring atoms like the atoms within the bulk of the nanowire. The unbonded atoms are often a source of defects within the nanowire, and may cause the nanowire to conduct electricity more poorly than the bulk material. As a nanowire shrinks in size, the surface atoms become more numerous compared to the atoms within the nanowire, and edge effects become more important.

Furthermore, the conductivity can undergo a quantization in energy: i.e. the energy of the electrons going through a nanowire can assume only discrete values, which are multiples of the conductance quantum $G = 2e^2/h$ (where e is the charge of the electron and h is the Planck constant).

The conductivity is hence described as the sum of the transport by separate *channels* of different quantized energy levels. The thinner the wire is, the smaller the number of channels available to the transport of electrons.

This quantization has been demonstrated by measuring the conductivity of a nanowire suspended between two electrodes while pulling it: as its diameter reduces, its conductivity decreases in a stepwise fashion and the plateaus correspond to multiples of G.

The quantization of conductivity is more pronounced in semiconductors like Si or GaAs than in metals, due to their lower electron density and lower effective mass. It can be observed in 25 nm wide silicon fins, and results in increased threshold voltage. In practical terms, this means that a MOSFET with such nanoscale silicon fins, when used in digital applications, will need a higher gate (control) voltage to switch the transistor on.

Welding Nanowires

To incorporate nanowire technology into industrial applications, researchers in 2008 developed a method of welding nanowires together: a sacrificial metal nanowire is placed adjacent to the ends of the pieces to be joined (using the manipulators of a scanning electron microscope); then an electric current is applied, which fuses the wire ends. The technique fuses wires as small as 10 nm.

For nanowires with diameters less than 10 nm, existing welding techniques, which require precise control of the heating mechanism and which may introduce the possibility of damage, will not be practical. Recently scientists discovered that single-crystalline ultrathin gold nanowires with diameters ~3–10 nm can be "cold-welded" together within seconds by mechanical contact alone, and under remarkably low applied pressures (unlike macro- and micro-scale cold welding process). High-resolution transmission electron microscopy and in situ measurements reveal that the welds are nearly perfect, with the same crystal orientation, strength and electrical conductivity as the rest of the nanowire. The high quality of the welds is attributed to the nanoscale sample dimensions, oriented-attachment mechanisms and mechanically assisted fast surface diffusion. Nanowire welds were also demonstrated between gold and silver, and silver nanowires (with diameters ~5–15 nm) at near room temperature, indicating that this technique may be generally applicable for ultrathin metallic nanowires. Combined with other nano- and microfabrication technologies, cold welding is anticipated to have potential applications in the future bottom-up assembly of metallic one-dimensional nanostructures.

Applications of Nanowires

Electronic Devices

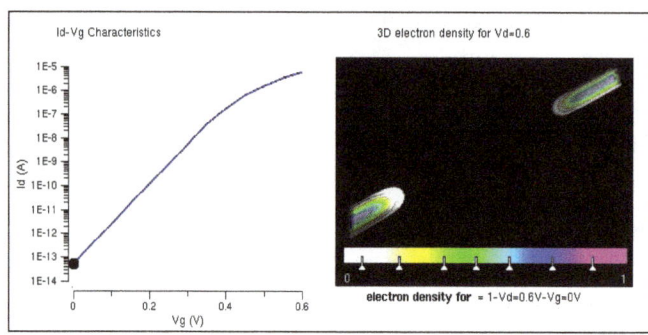

Atomistic simulation result for formation of inversion channel (electron density) and attainment of threshold voltage (IV) in a nanowire MOSFET. Note that the threshold voltage for this device lies around 0.45V.

Nanowires can be used for transistors. Transistors are used widely as fundamental building element in today's electronic circuits. As predicted by Moore's law, the dimension of transistors is shrinking smaller and smaller into nanoscale. One of the key challenges of building future nanoscale transistors is ensuring good gate control over the channel. Due to the high aspect ratio, if the gate dielectric is wrapped around the nanowire channel, we can get good control of channel electrostatic potential, thereby turning the transistor on and off efficiently.

To create active electronic elements, the first key step was to chemically dope a semiconductor nanowire. This has already been done to individual nanowires to create p-type and n-type semiconductors.

The next step was to find a way to create a p–n junction, one of the simplest electronic devices. This was achieved in two ways. The first way was to physically cross a p-type wire over an n-type wire. The second method involved chemically doping a single wire with different dopants along the length. This method created a p-n junction with only one wire.

After p-n junctions were built with nanowires, the next logical step was to build logic gates. By connecting several p-n junctions together, researchers have been able to create the basis of all logic circuits: the AND, OR, and NOT gates have all been built from semiconductor nanowire crossings.

In August 2012, researchers reported constructing the first NAND gate from undoped silicon nanowires. This avoids the problem of how to achieve precision doping of complementary nanocircuits, which is unsolved. They were able to control the Schottky barrier to achieve low-resistance contacts by placing a silicide layer in the metal-silicon interface.

It is possible that semiconductor nanowire crossings will be important to the future of digital computing. Though there are other uses for nanowires beyond these, the only ones that actually take advantage of physics in the nanometer regime are electronic.

In addition, nanowires are also being studied for use as photon ballistic waveguides as interconnects in quantum dot/quantum effect well photon logic arrays. Photons travel inside the tube, electrons travel on the outside shell.

When two nanowires acting as photon waveguides cross each other the juncture acts as a quantum dot.

Conducting nanowires offer the possibility of connecting molecular-scale entities in a molecular computer. Dispersions of conducting nanowires in different polymers are being investigated for use as transparent electrodes for flexible flat-screen displays.

Because of their high Young's moduli, their use in mechanically enhancing composites is being investigated. Because nanowires appear in bundles, they may be used as tribological additives to improve friction characteristics and reliability of electronic transducers and actuators.

Because of their high aspect ratio, nanowires are also uniquely suited to dielectrophoretic manipulation, which offers a low-cost, bottom-up approach to integrating suspended dielectric metal oxide nanowires in electronic devices such as UV, water vapor, and ethanol sensors.

Sensing of Proteins and Chemicals using Semiconductor Nanowires

In an analogous way to FET devices in which the modulation of conductance (flow of electrons/holes) in the semiconductor, between the input (source) and the output (drain) terminals, is controlled by electrostatic potential variation (gate-electrode) of the charge carriers in the device conduction channel, the methodology of a Bio/Chem-FET is based on the detection of the local change in charge density, or so-called "field effect", that characterizes the recognition event between a target molecule and the surface receptor.

This change in the surface potential influences the Chem-FET device exactly as a 'gate' voltage does, leading to a detectable and measurable change in the device conduction. When these devices are fabricated using semiconductor nanowires as the transistor element the binding of a chemical or biological species to the surface of the sensor can lead to the depletion or accumulation of charge carriers in the "bulk" of the nanometer diameter nanowire i.e. (small cross section available for conduction channels). Moreover, the wire, which serves as a tunable conducting channel, is in close contact with the sensing environment of the target, leading to a short response time, along with orders of magnitude increase in the sensitivity of the device as a result of the huge S/V ratio of the nanowires.

While several inorganic semiconducting materials such as Si, Ge, and metal oxides (e.g. In_2O_3, SnO_2, ZnO, etc.) have been used for the preparation of nanowires, Si is usually the material of choice when fabricating nanowire FET-based chemo/biosensors.

Several examples of the use of silicon nanowire(SiNW) sensing devices include the ultra sensitive, real-time sensing of biomarker proteins for cancer, detection of single virus particles, and the detection of nitro-aromatic explosive materials such as 2,4,6 Tri-nitrotoluene (TNT) in sensitives superior to these of canines. Silicon nanowires could also be used in their twisted form, as electromechanical devices, to measure intermolecular forces with great precision.

Limitations of Sensing with Silicon Nanowire FET Devices

Generally, the charges on dissolved molecules and macromolecules are screened by dissolved counterions, since in most cases molecules bound to the devices are separated from the sensor surface by approximately 2–12 nm (the size of the receptor proteins or DNA linkers bound to the sensor surface). As a result of the screening, the electrostatic potential that arises from charges on the analyte molecule decays exponentially toward zero with distance. Thus, for optimal sensing, the Debye length must be carefully selected for nanowire FET measurements. One approach of overcoming this limitation employs fragmentation of the antibody-capturing units and control over surface receptor density, allowing more intimate binding to the nanowire of the target protein. This approach proved useful for dramatically enhancing the sensitivity of cardiac biomarkers (e.g. Troponin) detection directly from serum for the diagnosis of acute myocardial infarction.

Silicon Nanowire

Silicon nanowires, also referred to as SiNWs, are a type of nanowire most often formed from a silicon precursor by etching of a solid or through catalyzed growth from a vapor or liquid phase. Initial synthesis is often accompanied by thermal oxidation steps to yield structures of accurately tailored size and morphology.

SiNWs have unique properties that are not seen in bulk (three-dimensional) Silicon materials. These properties arise from an unusual quasi one-dimensional electronic structure and are the subject of research across numerous disciplines and applications. The reason that SiNWs are considered as one of the most important one-dimensional materials is they could have a function as building blocks for nanoscale electronics assembled without the need for complex and costly fabrication facilities. SiNWs are frequently studied towards applications including photovoltaics, nanowire batteries, thermoelectrics and non-volatile memory.

Applications

Owing to their unique physical and chemical properties, silicon nanowires are a promising candidate for a wide range of applications that draw on their unique physico-chemical characteristics, which differ from those of bulk Silicon material.

SiNWs exhibit charge trapping behavior which renders such systems of value in applications necessitating electron hole separation such as photovoltaics, and photocatalysts.

Charge trapping behaviour and tuneable surface governed transport properties of SiNWs render this category of nanostructures of interest towards use as Metal Insulator Semiconductors and Field Effect Transistors, with further applications as nanoelectronic storage devices, in flash memory, logic devices as well as chemical and biological sensors.

The ability for lithium ions to intercalate into silicon structures renders various Si nanostructures of interest towards applications as anodes in Li-ion batteries (LiBs). SiNWs are of particular merit as such anodes as they exhibit the ability to undergo significant lithiation while maintaining structural integrity and electrical connectivity.

Synthesis

Several synthesis methods are known for SiNWs and these can be broadly divided into methods which start with bulk silicon and remove material to yield nanowires, also known as top-down synthesis, and methods which use a chemical or vapor precursor to build nanowires in a process generally considered to be bottom-up synthesis.

Top Down Synthesis Methods

These methods use material removal techniques to produce nanostructures from a bulk precursor:

- Laser beam ablation.
- Ion beam etching.
- Thermal evaporation oxide-assisted growth (OAG).

Bottom-up Synthesis Methods

- Vapour liquid solid (VLS) growth - a type of catalysed CVD often using silane as Si precursor and gold nanoparticles as catalyst (or 'seed').
- Molecular beam epitaxy - a form of PVD applied in plasma environment.
- Precipitation from a solution - A variation of the VLS method, aptly named Supercritical Fluid Liquid Solid (SFLS), that uses a supercritical fluid (e.g. organosilane at high temperature and pressure) as Si precursor instead of vapor. The catalyst would be a colloid in solution, such as colloidal gold nanoparticles, and the SiNWs are grown in this solution.

Thermal Oxidation of Silicon Nanowires

Subsequent to physical or chemical processing, either top-down or bottom-up, to obtain initial silicon nanostructures, thermal oxidation steps are often applied in order to obtain materials with desired size and aspect ratio. Silicon nanowires exhibit a distinct and useful self-limiting oxidation behaviour whereby oxidation effectively ceases due to diffusion limitations, which can be modeled. This phenomenon allows accurate control of dimensions and aspect rations in SiNWs and has been used to obtain high aspect ratio SiNWs with diameters below 5 nm. The self-limiting oxidation of SiNWs is of value towards lithium ion battery materials.

Orientation of Nanowires

The orientation of SiNWs has profound influence on the overall properties of systems. For this reason several procedures have been proposed for the alignment of nanowires in chosen orientations. This includes the use of electric fields in polar alignment, electrophoresis, mircofluidic methods and contact printing.

Outlook

There is significant interest in SiNWs for their unique properties and the ability to control size and aspect ratio with great accuracy. As yet, limitations in large-scale fabrication impede the uptake of this material in the full range of investigated applications. Combined studies of synthesis methods, oxidation kinetics and properties of SiNW systems aim to overcome the present limitations and facilitate the implementation of SiNW systems, for example, high quality vapor-liquid-solid–grown SiNWs with smooth surfaces can be reversibly stretched with 10% or more elastic strain, approaching the theoretical elastic limit of silicon, which could open the doors for the emerging "elastic strain engineering" and flexible bio-/nano-electronics.

Nanomesh

Perspective view of nanomesh, whose structure ends at the back of the figure. The distance between two pore centers is 3.2nm, and the pores are 0.05nm deep.

The nanomesh is a new inorganic nanostructured two-dimensional material, similar to graphene. It was discovered in 2003 at the University of Zurich, Switzerland.

It consists of a single layer of boron (B) and nitrogen (N) atoms, which forms by self-assembly a highly regular mesh after high-temperature exposure of a clean rhodium or ruthenium surface to borazine under ultra-high vacuum.

The nanomesh looks like an assembly of hexagonal pores at the nanometer (nm) scale. The distance between 2 pore centers is only of 3.2 nm, whereas each pore has a diameter of about 2 nm and is 0.05 nm deep. The lowest regions bind strongly to the underlying metal, while the wires (highest regions) are only bound to the surface through strong cohesive forces within the layer itself.

The boron nitride nanomesh is not only stable under vacuum, air and some liquids, but also up to temperatures of 796°C (1070 K). In addition it shows the extraordinary ability to trap molecules and metallic clusters, which have similar sizes to the nanomesh pores, forming a well-ordered array. These characteristics promise interesting applications of the nanomesh in areas like nano-catalysis, surface functionalisation, spintronics, quantum computing and data storage media like hard drives.

Structure

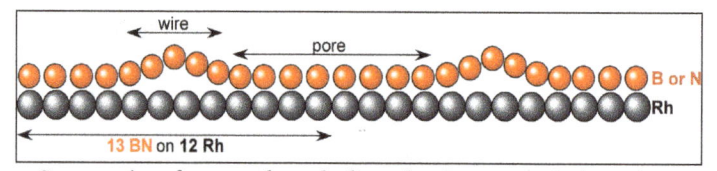

Cross-section of nanomesh on rhodium showing pore and wire regions.

h-BN nanomesh is a single sheet of hexagonal boron nitride, which forms on substrates like rhodium Rh(111) or ruthenium Ru(0001) crystals by a self-assembly process.

The unit cell of the h-BN nanomesh consists of 13x13 BN or 12x12 Rh atoms with a lattice constant of 3.2 nm. In a cross-section it means that 13 boron or nitrogen atoms are sitting on 12 rhodium atoms. This implies a modification of the relative positions of each BN towards the substrate atoms within a unit cell, where some bonds are more attractive or repulsive than other (site selective bonding), what induces the corrugation of the nanomesh.

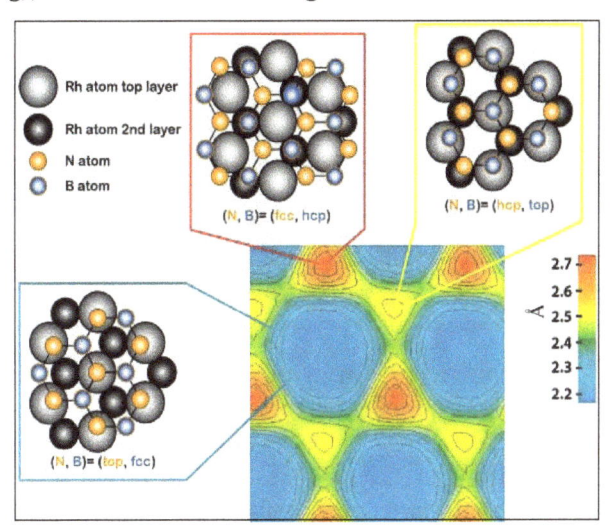

The image shows the boron nitride nanomesh measured by STM at 77K, where each "ball" represents one N atom. The center of each ring corresponds to the center of the pores.

The image is the theoretical calculation of the same area, where the N height relative to the underlying substrate is given.
The exact arrangement of Rh, N and B atoms is given for three different areas (blue: pores, yellow-red: wires).

The nanomesh corrugation amplitude of 0.05 nm causes a strong effect on the electronic structure, where two distinct BN regions are observed. They are easily recognized in the lower right image, which is a scanning tunneling microscopy (STM) measurement, as well as in the lower left image representing a theoretical calculation of the same area. A strongly bounded region assigned to the pores is visible in blue in the left image below (center of bright rings in the above image) and a weakly bound region assigned to the wires appears yellow-red in the left image below (area in-between rings in the above image).

Properties

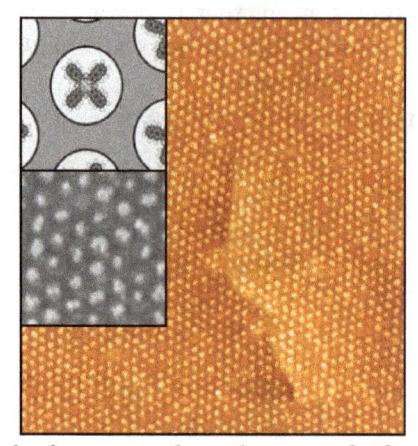

Naphthalocyanine molecules evaporated onto the nanomesh. They only adsorb in pores, forming a well-defined pattern.

The nanomesh is stable under a wide range of environments like air, water and electrolytes among others. It is also temperature resistant since it doesn't decompose up to 1275K under vacuum. In addition to these exceptional stabilities, the nanomesh shows the extraordinary ability to act as a scaffold for metallic nanoclusters and to trap molecules forming a well-ordered array.

In the case of gold (Au), its evaporation on the nanomesh leads to formation of well-defined round Au nanoparticles, which are centered at the nanomesh pores.

The STM figure shows Naphthalocyanine (Nc) molecules, which were vapor-deposited onto the nanomesh. These planar molecules have a diameter of about 2 nm, whose size is comparable to that of the nanomesh pores. It is spectacularly visible how the molecules form a well-ordered array with the periodicity of the nanomesh (3.22 nm). The lower inset shows a region of this substrate with higher resolution, where individual molecules are trapped inside the pores. In addition, the molecules seem to keep their native conformation, what means that their functionality is kept, which is nowadays a challenge in nanoscience.

Such systems with wide spacing between individual molecules/clusters and negligible intermolecular interactions might be interesting for applications such as molecular electronics and memory elements, in photochemistry or in optical devices.

Preparation and Analysis

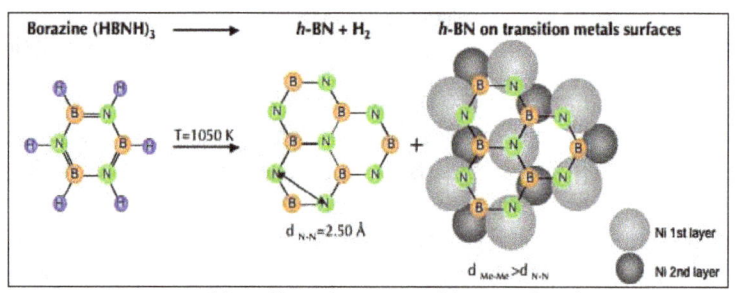

Decomposition of borazine on transition metal surfaces.

Well-ordered nanomeshes are grown by thermal decomposition of borazine $(HBNH)_3$, a colorless substance that is liquid at room temperature. The nanomesh results after exposing the atomically clean Rh(111) or Ru(0001) surface to borazine by chemical vapor deposition (CVD).

The substrate is kept at a temperature of 796 °C (1070 K) when borazine is introduced in the vacuum chamber at a dose of about 40 L (1 Langmuir = 10^{-6} torr sec). A typical borazine vapor pressure inside the ultrahigh vacuum chamber during the exposure is 3×10^{-7} mbar.

After cooling down to room temperature, the regular mesh structure is observed using different experimental techniques. Scanning tunneling microscopy (STM) gives a direct look on the local real space structure of the nanomesh, while low energy electron diffraction (LEED) gives information about the surface structures ordered over the whole sample. Ultraviolet photoelectron spectroscopy (UPS) gives information about the electronic states in the outermost atomic layers of a sample, i.e. electronic information of the top substrate layers and the nanomesh.

Nanohole

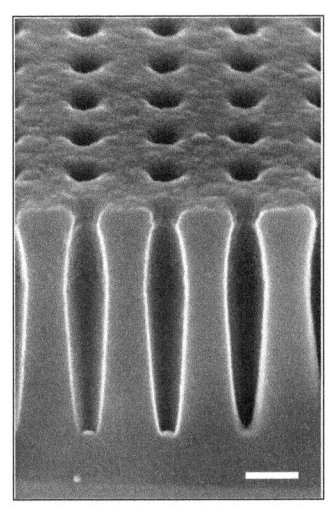

Angled cross-sectional scanning electron micrograph of a nanohole array etched in amorphous silicon, with a thin conductive polymer coating. Scale bar is 200 nm.

Nanoholes are a class of nanostructured material consisting of nanoscale voids in a surface of a material. It is different than nanofoam or nanoporous materials which support a network of voids permeating throughout the material (often in a disordered state), nanohole materials feature a regular hole pattern extending through a single surface. These can be thought of as the inverse of a nanopillar or nanowire structure.

Uses

Nanohole structures have been used for a variety of applications, ranging from superlenses produced from a metal nanohole array, to structured photovoltaic devices used to improve carrier extraction, and light absorption.

Nanohole structures are also extensively utilized for the creation of photonic crystals, particularly for creating photonic crystal waveguides.

Nanosheet

A nanosheet is a two-dimensional nanostructure with thickness in a scale ranging from 1 to 100 nm.

A typical example of a nanosheet is graphene, the thinnest two-dimensional material (0.34 nm) in the world. It consists of a single layer of carbon atoms with hexagonal lattices.

Examples and Applications

As of 2017 Silicon nanosheets are being used to prototype future generations of small (5 nm) transistors.

Carbon nanosheets (from hemp) may be an alternative to graphene as electrodes in supercapacitors.

Synthesis

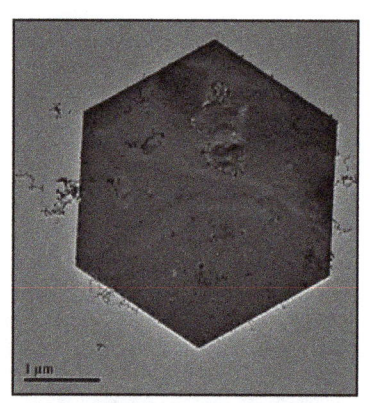

TEM image of PbO nanosheets with highly symmetric edge length. The edges of PbO nanosheets are surrounded with Au NPs seeds.

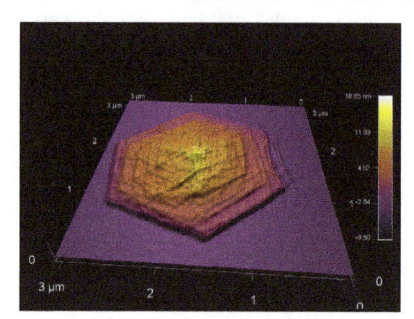

3D AFM topography image of multilayered palladium nanosheet on silicon wafer.

The most commonly used nanosheet synthesis methods use a bottom-up approach, e.g., pre-organization and polymerization at interfaces like Langmuir–Blodgett films, solution phase synthesis and chemical vapor deposition (CVD). For example, CdTe (cadmium telluride) nanosheets could be synthesized by precipitating and aging CdTe nanoparticles in deionized water. The formation of free-floating CdTe nanosheets was due to directional hydrophobic attraction and anisotropic electrostatic interactions caused by dipole moment and small positive charges. Molecular simulations through a coarse-grained model with parameters from semi-empirical quantum mechanics calculations can be used to prove the experimental process.

Ultrathin single-crystal PbS (lead sulfur) sheets with micro scale in x-, y- dimensions can be obtained using a hot colloidal synthesis method. Compounds with linear chloroalkanes like 1,2-dichloroethane containing chlorine were used during the formation of PbS sheets. PbS ultrathin sheets probably resulted from the oriented attachment of the PbS nanoparticles in a two-dimensional fashion. The highly reactive facets were preferentially consumed in the growth process that led to the sheet-like PbS crystal growth.

Nanosheets can also be prepared at room temperature. For instance, hexagonal PbO (lead oxide)) nanosheets were synthesized using gold nanoparticles as seeds under room temperature. The size of the PbO nanosheet can be tuned by gold NPs and Pb^{2+} concentration in the growth solution. No organic surfactants were employed in the synthesis process. Oriented attachment, in which the sheets form by aggregation of small nanoparticles that each has a net dipole moment, and ostwald ripening are the two main reasons for the formation of the PbO nanosheets.

Carbon nanosheets have been produced using industrial hemp bast fibres with a technique that involves heating the fibres at over 350F (180C) for 24 hours. The result is then subjected to intense heat causing the fibers to exfoliate into a carbon nanosheet. This has been used to create an electrode for a supercapacitor with electrochemical qualities 'on a par with' devices made using graphene.

Metal nanosheets have also been synthesized from solution-based method by reducing metal precursors, including palladium, rhodium, and gold.

Nanopillar

Nanopillars is an emerging technology within the field of nanostructures. Nanopillars are pillar shaped nanostructures approximately 10 nanometers in diameter that can be grouped together in lattice like arrays. They are a type of metamaterial, which means that nanopillars get their attributes from being grouped into artificially designed structures and not their natural properties. Nanopillars set themselves apart from other nanostructures due to their unique shape. Each nanopillar has a pillar shape at the bottom and a tapered pointy end on top. This shape in combination with nanopillars' ability to be grouped together exhibits many useful properties. Nanopillars have many applications including efficient solar panels, high resolution analysis, and antibacterial surfaces.

Applications

Solar Panels

Due to their tapered ends, nanopillars are very efficient at capturing light. Solar collector surfaces coated with nanopillars are three times as efficient as nanowire solar cells. Less material is needed to build a solar cell out of nanopillars compared to regular semi conductive materials.

They also hold up well during the manufacturing process of solar panels. This durability allows manufacturers to use cheaper materials and less expensive methods to produce solar panels. Researchers are looking into putting dopants into the bottom of the nanopillars, to increase the amount of time photons will bounce around the pillars and thus the amount of light captured. As well as capturing light more efficiently, using nanopillars in solar panels will allow them to be flexible. The flexibility gives manufacturers more options on how they want their solar panels to be shaped as well as reduces costs in terms of how delicately the panels have to be handled. Although nanopillars are more efficient and cheaper than standard materials, scientists have not been able to mass-produce them yet. This is a significant drawback to using nanopillars as a part of the manufacturing process.

Antibacterial Surfaces

Nanopillars also have functions outside of electronics and can imitate natures defenses. Cicadas' wings are covered in tiny, nanopillar shaped rods. When bacteria hits a cicada's wing, it gets caught on the rods. The rods don't hold the bacteria up evenly so they get punctured and die. Since the rods on the cicadas are about the same size and shape as artificial nanopillars, it is possible for humans to copy this defense. A surface covered with nanopillars would immediately become germ free. If mass-produced and installed everywhere, nanopillars could eliminate much of the risk of transmitting diseases through touching infected surfaces.

High Resolution Molecular Analysis

Another use of nanopillars is observing cells. Nanopillars capture light so well that when lights hits them, the glow the nanopillars emit dies down at around 150 nanometers. Because this distance is less than the wavelength of light, it allows researchers to observe small objects without the interference of background light. This is especially useful in cellular analysis. The cells group around the nanopillars because of its small size and recognize it as an organelle. The nanopillars simply hold the cells in place while the cells are being observed.

History

In 2006, researchers at the University of Nebraska-Lincoln and the Lawrence Livermore National Laboratory developed a cheaper and more efficient way to create nanopillars. They used a combination of nanosphere lithography (a way of organizing the lattice) and reactive ion etching(molding the nanopillars to the right shape) to make large groups of silicon pillars with less than 500 nm diameters. Then, in 2010, researchers fabricated a way to manufacture nanopillars with tapered ends. The former design of a pillar with a flat blunt top reflected much of the light coming onto the pillars. The tapered tops allow light to enter the forest of nanopillars and the wider bottom absorbs almost all of the light that hits it. This design captures about 99% of the light whereas nanorods which have a uniform thickness only captured 85% of the light. After the introduction of tapered ends, researchers started to find many more applications for nanopillars.

Manufacturing Process

Constructing nanopillars is a simple but lengthy procedure that can take hours. The process to create nanopillars starts with anodizing a 2.5 mm thick aluminum foil mold. Anodizing the foil creates pores in the foil a micrometer deep and 60 nanometers wide. The next step is to treat the foil with phosphoric acid which expands the pores to 130 nanometers. The foil is anodized once more making its pores a micrometer deeper. Lastly, a small amount of gold is added to the pores to catalyze the reaction for the growth of the semiconductor material. When the aluminum is scraped away there is a forest of nanopillars left inside a casing of aluminum oxide.

Nanostructured Film

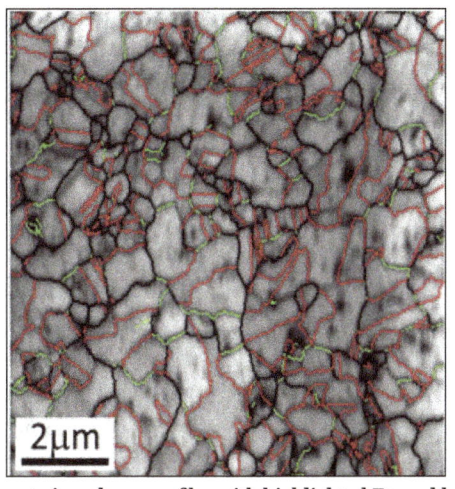

Surface of a nanotwinned copper film with highlighted Σ3 and low angle grain boundaries as imaged by EBSD. Image adapted from Zhao et al.

A nanostructured film is a film resulting from engineering of nanoscale features, such as dislocations, grain boundaries, defects, or twinning. In contrast to other nanostructures, such as nanoparticles, the film itself may be up to several microns thick, but possesses a large concentration of nanoscale features homogeneously distributed throughout the film. Like other nanomaterials, nanostructured films have sparked much interest as they possess unique properties not found in bulk, non-nanostructured material of the same composition. In particular, nanostructured films have been the subject of recent research due to their superior mechanical properties, including strength, hardness, and corrosion resistance compared to regular films of the same material. Examples of nanostructured films include those produced by grain boundary engineering, such as nano-twinned ultra-fine grain copper, or dual phase nanostructuring, such as crystalline metal and amorphous metallic glass nanocomposites.

Synthesis and Characterization

Nanostructured films are commonly created using magnetron sputtering from an appropriate target material. Films can be elemental in nature, formed by sputtering from a pure metal target such as copper, or composed of compound materials. Varying parameters such as the sputtering rate, substrate temperature, and sputtering interrupts allow the creation of films with a variety of different nanostructured elements. Control over nano-twinning, tailoring of specific types of grain boundaries, and restricting the movement and propagation of dislocations have been demonstrated using films produced via magnetron sputtering.

Methods used to characterize nanostructured films include transmission electron microscopy, scanning electron microscopy, electron backscatter diffraction, focused ion beam milling, and nanoindentation. These techniques are used as they allow imaging of nanoscale structures, including dislocations, twinning, grain boundaries, film morphology, and atomic structure.

Material Properties

Nanostructured films are of interest due to their superior mechanical and physical properties compared to their normal equivalent. Elemental nanostructured films composed of pure copper were found to possess good thermal stability due to the nano-twinned film possessing a higher fraction of grain boundaries. In addition to possessing higher thermal stability, copper films that were highly nano-twinned were found to have a better corrosion resistance than copper films with a low concentration of nano-twins. Control of the fraction of grains in a material with nano-twins present has great potential for less expensive alloys and coatings with a good degree of corrosion resistance.

Compound nanostructured films composed of crystalline $MgCu_2$ cores encapsulated by amorphous glassy shells of the same material were shown to possess a near-ideal mechanical strength. The crystalline $MgCu_2$ cores, typically less than 10 nm in size, were found to substantially strengthen the material by restricting the movement of dislocations and grains. The cores were also found to contribute to overall material strength by restricting the movement of shear bands in the material. This nanostructured film differs from both crystalline metals and amorphous metallic glasses, both of which exhibit behaviors such as the reverse Hall-Petch and shear-band softening effects that prevent them from reaching ideal strength values.

Applications

Nanostructured films with superior mechanical properties allow previously unusable materials to be utilized in new applications, enabling advances fields where coatings are heavily utilized, such as aerospace, energy, and other engineering fields. Production scalability of nanostructured films has already been demonstrated, and the ubiquity of sputtering techniques in industry is predicted to facilitate the incorporation of nanostructured films into existing applications.

Sculptured Thin Film

Sculptured thin films (STFs) are nanostructured materials with unidirectionally varying properties that can be designed and realized in a controllable manner using variants of physical vapor deposition. The ability to virtually instantaneously change the growth direction of their columnar morphology, through simple variations in the direction of the incident vapor flux, leads to a wide spectrum of columnar forms.

Forms

These forms can be:
1. two-dimensional, ranging from the simple slanted columns and chevrons to the more complex C- and S-shaped morphologies.

2. three-dimensional, including simple helixes and superhelixes.

3. combinations of two- and three-dimensional forms.

Properties

The column diameter and the column separation normal to the thickness direction of any STF are nominally constant. The column diameter can range from about 10 to 300 nm, while the density may lie between its theoretical maximum value to less than 20% thereof. The crystallinity must be at a scale smaller than the column diameter. The chemical composition is essentially unlimited, ranging from insulators to semiconductors to metals. Most recently, polymeric STFs have been deposited by combining physical and chemical vapor deposition processes; and deposition on micropatterned substrates has also been carried out.

Uses

To date, the chief applications of STFs are in optics as polarization filters, Bragg filters, and spectral hole filters. At visible and infrared wavelengths, a single-section STF is a unidirectionally nonhomogeneous continuum with direction-dependent properties. Several sections can be grown consecutively into a multisection STF, which can be conceived of as an optical circuit that can be integrated with electronic circuitry on a chip. Being porous, an STF can act as a sensor of fluids and can be impregnated with liquid crystals for switching applications too. Applications as low-permittivity barrier layers in electronic chips as well as solar cells have also been suggested. Biomedical applications such as tissue scaffolds, drug-delivery platforms, virus traps, and labs-on-a-chip are also in different stages of development.

Gradient Multilayer Nanofilm

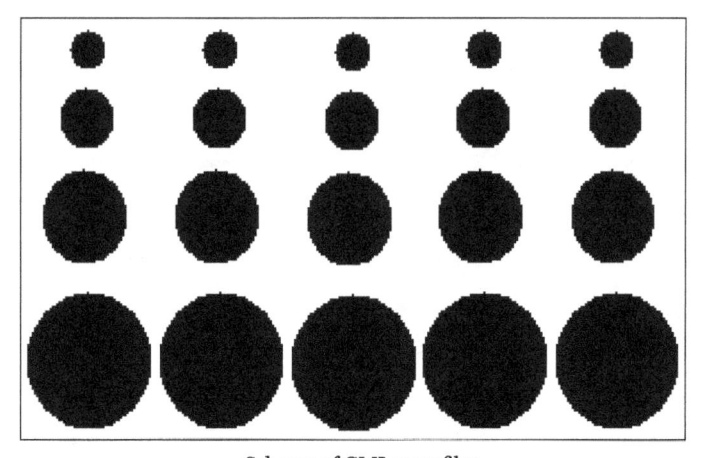

Schema of GML nanofilm.

Gradient multilayer (GML) nanofilm is an assembly of quantum dot layers with a built-in gradient of nanoparticle size, composition or density.

Properties of such nanostructure are finding its applications in design of solar cells and energy storage devices.

The GML nanostructure can be embedded in the organic material (polymer), or can include quantum dots made of two or more types of material.

Photovoltaic Applications

The GML nanofilm only 100 nanometers thick can absorb the entire Sun spectrum (0.3–2.0+ eV). At the same time, gradient of the quantum dots size can create a gradient of the electrochemical potential, acting as an equivalent of built-in electric field inside a nanofilm. This enhances transport of electrons and holes, and improves internal quantum efficiency (IQE) and photocurrent.

Manufacturing

The industrial manufacturing of GML nanofilms represents a challenge. Traditional methods of building nanostructured materials (like spin coating) can't form GML nanostructures, while more effective methods like Atomic Layer Deposition (ALD) or Langmuir-Blodget "microchemical" method. are expensive.

Icosahedral Twins

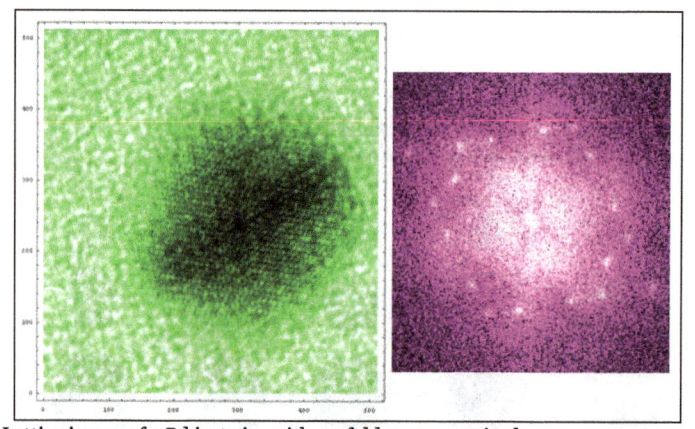

Lattice image of a Pd icotwin, with 10-fold symmetry in the power spectrum.

An icosahedral twin is a twenty-face cluster made of ten interlinked dual-tetrahedron (bowtie) crystals, typically joined along triangular (e.g. cubic-111) faces having three-fold symmetry. One can think of their formation as a kind of nano-scale self-assembly.

A variety of nanostructures (e.g. condensing argon, metal atoms, and virus capsids) assume icosahedral form on size scales where surface forces eclipse those from the bulk. A twinned-form of these nanostructures is sometimes found to occur e.g. in face-centered-cubic (FCC) metal-atom clusters larger than 10 nm in diameter. This may occur when the building-blocks beneath each of the 20 facets of an initially icosahedral cluster make the case for conversion to a translationally symmetric crystalline form.

Causes

FCC icotwin model down 5-fold & 3-fold zones.

When interatom bonding does not have strong directional preferences, it is not unusual for atoms to gravitate toward a kissing number of 12 nearest neighbors. The three most symmetric ways to do this are by icosahedral clustering, or by crystalline face-centered-cubic (cuboctahedral) and/or hexagonal (tri-orthobicupolar) close packing.

Icosahedral arrangements, perhaps because of their slightly smaller surface area, may be preferred for small clusters e.g. noble gas and metal atoms in condensed phases (both liquid and solid). However, the Achilles heel for icosahedral clustering about a single point is that it cannot fill space over large distances in a way that is translationally ordered.

Hence bulk atoms (i.e. sufficiently large clusters) generally revert to one of the crystalline close-packing configurations instead. In other words, when icosahedral clusters get sufficiently large, the bulk-atom vote wins out over the surface-atom vote, and the atoms beneath each of the 20 facets adopt a face-centered-cubic pyramidal arrangement with tetrahedral (111) facets. Thus icosahedral twins are born, with a certain amount of strain along the interfacial (111) planes.

Ubiquity

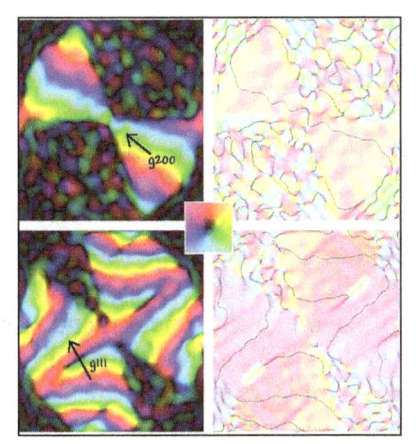

Darkfield analysis of dual-tetrahedron crystal pairs.

Icosahedral-twinning has been seen in face-centered-cubic metal-nanoparticles that have nucleated: (i) by evaporation onto surfaces, (ii) out of solution, and (iii) by reduction in a polymer matrix. Icosahedral (or decahedral) twinning seems to be the result when clusters growing outward from an icoashedral-seed get larger than about 10[nm] in diameter.

Quasicrystals are *un-twinned* structures with long range rotational but not translational periodicity, that some initially tried to explain away as icosahedral twinning. Quasi-crystals let non-space-filling coordination persist to larger size scales. However, they generally form only when the compositional makeup (e.g. of two dissimilar metals like Ti and Mn) serves as an antagonist to formation of one of the more common close-packed space-filling but twinned crystalline forms.

Application

Face-centered-cubic noble-metal atom-clusters are important chemical-reaction nano-catalysts. One example of this is the platinum used in automobile catalytic converters. Icosahedral twinning makes it possible to cover the entire surface of a nanoparticle with {111} facets, in cases where those particular atomic-facets show favorable catalytic-activity.

Detection

Electron diffraction and high-resolution transmission electron microscope (TEM) imaging are two methods for identifying the icosahedral-twin structure of *individual* clusters. Digital darkfield analysis of lattice-fringe images shows promise for recognition of icosahedral twinning from most of the randomly oriented clusters in a microscope-image field of view.

Nano Flake

In a general meaning a nanoflake is a flake (that is, an uneven piece of material with one dimension substantially smaller than the other two) with at least one nanometric dimension (that is, between 1 and 100 nm). A flake is not necessarily perfectly flat but it is characterized by a plate-like form or structure. There are nanoflakes of all sorts of materials.

In a more restricted meaning, in the context of solar energy, Nano Flakes are a type of semiconductor that has potential for solar energy creation as the product itself is only in the prototype phase. With its crystalline structure the crystals are able to absorb light and harvest 30 percent of solar energy directed at its surface. These Nano Flakes can potentially also help out with economic and environmental problems associated with solar energy. Working on this Nano Flake Technology is Dr. Martin Aagesen at the Niels Bohr Institute at the University of Copenhagen with a PhD from the Nano Science Center. When Dr. Martin Aagesen discovered and published this new idea in 2007 there was much publicity about it and sparked many people to work on it. One major company that is working on these Nano structures is SunFlake which is a science company from the Nano Science Center with Dr. Martin Aagesen as Chief Executive Officer (CEO). The funding for the Nano flakes came from Danish Venture Capital fund SEED capital and University of Copenhagen.

Structure

The Nano flakes have a structure that contains tiny crystals in which millions of these crystals could fit into a single square centimeter. The tiny crystals absorb the sunlight and use the solar energy to convert it to electricity. This perfect crystalline structure is why this product can revolutionize solar energy. The large surface to volume ratio and the texture of the surface of this nano structure provides a larger absorption rate of the sun's light energy. Also researchers are working on trying to combine it with different semiconducting materials since the usual requirements of a need for a similar crystal structure for the carrier substrate is less stressed in the Nano flakes structure. The carrier substrate in the Nano flakes purpose is to permit growth of the nano structures and works as a contact for the Nano structures when they are actively absorbing the sun's energy.

Purpose

Solar energy obtained from the Nano flakes can help benefit in a couple of ways. Nano flakes can potentially help lower the cost of solar energy. Also since more solar energy can theoretically be obtained from Nano flakes, their use can potentially will keep the earth's environment cleaner by reducing the need for fossil fuels.

Cost

The high cost of solar energy stems from the difficulty of converting the solar energy into electricity for use, and less than 1 percent of the world's electricity comes from the sun because of this process. Nano flakes can potentially help with the economic issues of solar energy by lowering the cost due to an easier process and a better outcome of energy. Nano flake technology can potentially make it easier to convert solar energy into electricity estimated at twice the amount that today's solar cells can harvest. This new technology can also potentially lower the cost of solar energy because it allows for a reduction in expensive semiconducting silicon. Energy loss is also potentially reduced with a shorter distance of the solar energy transportation across smaller Nano flakes.

Environment

Nano Flake technology can also help keep the environment cleaner as the sun as the source it produces clean pure sustainable energy that can be converted into electricity. While fossil fuel is the primary energy source for electricity, using solar energy obtained from Nano flakes will lower dependence on fossil fuels. When fossil fuels are burned for use they release a toxic gas which has a huge impact on earth's pollution. Also the process of obtaining these fossil fuels is not good for the environment, whether it be mining for coal, drilling for oil, and hydraulic fracturing the earth's surface to reach the oil and gas.

Nanofoam

Nanofoams are a class of nanostructured, porous materials (foams) containing a significant population of pores with diameters less than 100 nm. Aerogels are one example of nanofoam.

Metal

In 2006, researchers produced metal nanofoams by igniting pellets of energetic metal bis(tetrazolato)amine complexes. Nanofoams of iron, cobalt, nickel, copper, silver, and palladium have been prepared through this technique. These materials exhibit densities as low as 11 mg/cm^3, and surface areas as high as 258 m^2/g. These foams are effective catalysts.

Carbon

Carbon nanofoam is an allotrope of carbon discovered in 1997. It consists of a cluster-assembly of carbon atoms strung together in a loose three-dimensional web. The material has a density of 2–10 mg/cm^3 (0.0012 lb/ft^3).

Glass

In 2014, researchers also fabricated glass nanofoam via femtosecond laser ablation. Their work consisted of raster scanning femtosecond laser pulses over the surface of glass to produce glass nanofoam with ~70 nm diameter wires.

Thermodynamics of Nanostructures

As the devices continue to shrink further into the sub-100 nm range following the trend predicted by Moore's law, the topic of thermal properties and transport in such nanoscale devices becomes increasingly important. Display of great potential by nanostructures for thermoelectric applications also motivates the studies of thermal transport in such devices. These fields, however, generate two contradictory demands: high thermal conductivity to deal with heating issues in sub-100 nm devices and low thermal conductivity for thermoelectric applications. These issues can be addressed with phonon engineering, once nanoscale thermal behaviors have been studied and understood.

The Effect of the Limited Length of Structure

In general two carrier types can contribute to thermal conductivity - electrons and phonons. In nanostructures phonons usually dominate and the phonon properties of the structure become of a particular importance for thermal conductivity. These phonon properties include: phonon group velocity, phonon scattering mechanisms, heat capacity, Grüneisen parameter. Unlike bulk materials, nanoscale devices have thermal properties which are complicated by boundary effects due to small size. It has been shown that in some cases phonon-boundary scattering effects dominate the thermal conduction processes, reducing thermal conductivity.

Depending on the nanostructure size, the phonon mean free path values (Λ) may be comparable or larger than the object size, L. When L is larger than the phonon mean free path, Umklapp scattering process limits thermal conductivity (regime of diffusive thermal conductivity). When L is comparable to or smaller than the mean free path (which is of the order 1 µm for carbon nanostructures), the continuous energy model used for bulk materials no longer applies and nonlocal and nonequilibrium aspects to heat transfer also need to be considered. In this case phonons in defectless structure could propagate without scattering and thermal conductivity becomes ballistic (similar to ballistic conductivity). More severe changes in thermal behavior are observed when the feature size L shrinks further down to the wavelength of phonons.

Nanowires

Thermal Conductivity Measurements

The first measurement of thermal conductivity in silicon nanowires was published in 2003. Two important features were pointed out: 1) The measured thermal conductivities are significantly lower than that of the bulk Si and, as the wire diameter is decreased, the corresponding thermal conductivity is reduced. 2) As the wire diameter is reduced, the phonon boundary scattering dominates over phonon–phonon Umklapp scattering, which decreases the thermal conductivity with an increase in temperature.

For 56 nm and 115 nm wires $k \sim T^3$ dependence was observed, while for 37 nm wire $k \sim T^2$ dependence and for 22 nm wire $k \sim T$ dependence were observed. Chen et al. has shown that the one-dimensional cross-over for 20 nm Si nanowire occurs around 8K, while the phenomenon was observed for temperature values greater than 20K. Therefore, the reason of such behaviour is not in the confinement experienced by phonons so that three-dimensional structures display two-dimensional or one-dimensional behavior.

Theoretical Models for Nanowires

Different Phonon Modes Contribution to Thermal Conductivity

Assuming that Boltzmann transport equation is valid, thermal conductivity can be written as:

$$k = \frac{1}{3}Cv_g\Lambda = \frac{1}{3}Cv_g^2\tau$$

where C is the heat capacity, v_g is the group velocity and τ is the relaxation time. Note that this assumption breaks down when the dimensions of the system are comparable to or smaller than the wavelength of the phonons responsible for thermal transport. In our case, phonon wavelengths are generally in the 1 nm range and the nanowires under consideration are within tens of nanometers range, the assumption is valid.

Different phonon mode contributions to heat conduction can be extracted from analysis of the experimental data for silicon nanowires of different diameters to extract the $C \cdot v_g$ product for analysis. It was shown that all phonon modes contributing to thermal transport are excited well below the Si Debye temperature (645 K).

From the thermal conductivity equation, one can write the product $C \cdot v_g$ for each isotropic phonon branch i.

$$(Cv_g)_i = \frac{k_B^4 T^3}{2\hbar^3 \pi^2} \int \frac{1}{v_{p,i}^2} \left[\frac{x^4 \exp(x)}{(\exp(x)-1)^2} \right] dx$$

where $x = \hbar\omega/k_BT$ and $v_{p,i}$ is the phonon phase velocity, which is less sensitive to phonon dispersions than the group velocity v_g.

Many models of phonon thermal transport ignores the effects of transverse acoustic phonons (TA) at high frequency due to their small group velocity. (Optical phonon contributions are also ignored for the same reason.) However, upper branch of TA phonons have non-zero group velocity at the Brillouin zone boundary along the Γ-K direction and, in fact, behave similarly to the longitudinal acoustic phonons (LA) and can contribute to the heat transport.

Then, the possible phonon modes contributing to heat conduction are both LA and TA phonons at low and high frequencies. Using the corresponding dispersion curves, the $C \cdot v_g$ product can then be calculated and fitted to the experimental data. The best fit was found when contribution of high-frequency TA phonons is accounted as 70% of the product at room temperature. The remaining 30% is contributed by the LA and TA phonons at low-frequency.

Using Complete Phonon Dispersions

Thermal conductivity in nanowires can be computed based on complete phonon dispersions instead of the linearlized dispersion relations commonly used to calculate thermal conductivity in bulk materials.

Assuming the phonon transport is diffusive and Boltzmann transport equation (BTE) is valid, nanowire thermal conductance $G(T)$ can be defined as:

$$G(T) \simeq \sum_\alpha \int \frac{\lambda_\alpha(k_z)}{L} \frac{\hbar\omega_\alpha(k_z)}{2\pi} \frac{df_b}{dT} v_z(\alpha, k_z) dk_z$$

where the variable α represents discrete quantum numbers associated with sub-bands found in one-dimensional phonon dispersion relations, f_B represents the Bose-Einstein distribution, v_z is the phonon velocity in the z direction and λ is the phonon relaxation length along the direction of the wire length. Thermal conductivity is then expressed as:

$$k(T) = \frac{1}{S}\sum \alpha \int_0^{\frac{\pi}{a_z}} \lambda_\alpha(k_z) \frac{\hbar\omega_\alpha(k_z)}{2\pi} \frac{df_B}{dT} v_z(\alpha, k_z) dk_z$$

where S is the cross sectional area of the wire, a_z is the lattice constant.

It was shown that, using this formula and atomistically computed phonon dispersions (with interatomic potentials developed in), it is possible to predictively calculate lattice thermal conductivity curves for nanowires, in good agreement with experiments. On the other hand, it was not possible to obtain correct results with the approximated Callaway formula. These results are expected to apply to "nanowhiskers" for which phonon confinement effects are unimportant. Si nanowires wider than ~35 nm are within this category.

Very Thin Nanowires

For large diameter nanowires, theoretical models assuming the nanowire diameters are comparable to the mean free path and that the mean free path is independent of phonon frequency have been able to closely match the experimental results. But for very thin nanowires whose dimensions are comparable to the dominant phonon wavelength, a new model is required. The study in has shown that in such cases, the phonon-boundary scattering is dependent on frequency. The new mean free path is then should be used:

$$l^{-1} = B\left(\frac{h}{d}\right)^2 \frac{1}{d}\left(\frac{\omega}{\omega_D}\right)^2 N(\omega)$$

Here, l is the mean free path (same as Λ). The parameter h is length scale associated with the disordered region, d is the diameter, $N(\omega)$ is number of modes at frequency ω, and B is a constant related to the disorder region.

Thermal conductance is then calculated using Landauer formula:

$$G(T) = \frac{1}{2\pi\hbar}\int_0^\infty \left(\frac{N_1(\omega)}{1+L/l(\omega)} + \frac{N_2(\omega)}{1+L/d}\right)\frac{\hbar^3\omega^2}{k_B T^2} \times \frac{e^{\frac{\hbar\omega}{k_B T}}}{(e^{\frac{\hbar\omega}{k_B T}} - 1)^2}$$

Carbon Nanotubes

As nanoscale graphitic structures, carbon nanotubes are of great interest for their thermal properties. The low-temperature specific heat and thermal conductivity show direct evidence of 1-D quantization of the phonon band structure. Modeling of the low-temperature specific heat allows determination of the on-tube phonon velocity, the splitting of phonon subbands on a single tube, and the interaction between neighboring tubes in a bundle.

Thermal Conductivity Measurements

Measurements show a single-wall carbon nanotubes (SWNTs) room-temperature thermal conductivity about 3500 W/(m·K), and over 3000 W/(m·K) for individual multiwalled carbon nanotubes (MWNTs). It is difficult to replicate these properties on the macroscale due to imperfect contact between individual CNTs, and so tangible objects from CNTs such as films or fibres have reached only up to 1500 W/(m·K) so far. Addition of nanotubes to epoxy resin can double the thermal conductivity for a loading of only 1%, showing that nanotube composite materials may be useful for thermal management applications.

Theoretical Models for Nanotubes

Thermal conductivity in CNT is mainly due to phonons rather than electrons so the Wiedemann–Franz law is not applicable.

In general, the thermal conductivity is a tensor quality, but for this discussion, it is only important to consider the diagonal elements:

$$k_{zz} = \sum C v_z^2 \tau$$

where C is the specific heat, and v_z and τ are the group velocity and relaxation time of a given phonon state.

At low temperatures (T is far less than Debye temperature), the relaxation time is determined by scattering of fixed impurities, defects, sample boundaries, etc. and is roughly constant. Therefore, in ordinary materials, the low-temperature thermal conductivity has the same temperature dependence as the specific heat. However, in anisotropic materials, this relationship does not strictly hold. Because the contribution of each state is weighted by the scattering time and the square of the velocity, the thermal conductivity preferentially samples states with large velocity and scattering time. For instance, in graphite, the thermal conductivity parallel to the basal planes is only weakly dependent on the interlayer phonons. In SWNT bundles, it is likely that $k(T)$ depends only on the on-tube phonons, rather than the intertube modes.

Thermal conductivity is of particular interest in low-dimensional systems. For CNT, represented as 1-D ballistic electronic channel, the electronic conductance is quantized, with a universal value of:

$$G_0 = \frac{2e^2}{h}$$

Similarly, for a single ballistic 1-D channel, the thermal conductance is independent of materials parameters, and there exists a quantum of thermal conductance, which is linear in temperature:

$$G_{th} = \frac{\pi^2 k_B^2 T}{3h}$$

Possible conditions for observation of this quantum were examined by Rego and Kirczenow. In 1999, Keith Schwab, Erik Henriksen, John Worlock, and Michael Roukes carried out a series of experimental measurements that enabled first observation of the thermal conductance quantum. The measurements employed suspended nanostructures coupled to sensitive dc SQUID measurement devices. In 2008, a colorized electron micrograph of one of the Caltech devices was acquired for the permanent collection of the Museum of Modern Art in New York.

At high temperatures, three-phonon Umklapp scattering begins to limit the phonon relaxation time. Therefore, the phonon thermal conductivity displays a peak and decreases with increasing temperature. Umklapp scattering requires production of a phonon beyond the Brillouin zone boundary; because of the high Debye temperature of diamond and graphite, the peak in the thermal conductivity of these materials is near 100 K, significantly higher than for most other materials. In less crystalline forms of graphite, such as carbon fibers, the peak in $k(T)$ occurs at higher temperatures, because defect scattering remains dominant over Umklapp scattering to higher temperature. In low-dimensional systems, it is difficult to conserve both energy and momentum for Umklapp processes, and so it may be possible that Umklapp scattering is suppressed in nanotubes relative to 2-D or 3-D forms of carbon.

Berber *et al.* have calculated the phonon thermal conductivity of isolated nanotubes. The value $k(T)$ peaks near 100 K, and then decreases with increasing temperature. The value of $k(T)$ at the peak (37,000 W/(m·K)) is comparable to the highest thermal conductivity ever measured (41,000

W/(m·K) for an isotopically pure diamond sample at 104 K). Even at room temperature, the thermal conductivity is quite high (6600 W/(m·K)), exceeding the reported room-temperature thermal conductivity of isotopically pure diamond by almost a factor of 2.

In graphite, the interlayer interactions quench the thermal conductivity by nearly 1 order of magnitude. It is likely that the same process occurs in nanotube bundles. Thus it is significant that the coupling between tubes in bundles is weaker than expected. It may be that this weak coupling, which is problematic for mechanical applications of nanotubes, is an advantage for thermal applications.

Phonon Density of States for Nanotubes

The phonon density of states is to calculated through band structure of isolated nanotubes, which is studied in Saito *et al.* and Sanchez-Portal *et al.* When a graphene sheet is "rolled" into a nanotube, the 2-D band structure folds into a large number of 1-D subbands. In a (10,10) tube, for instance, the six phonon bands (three acoustic and three optical) of the graphene sheet become 66 separate 1-D subbands. A direct result of this folding is that the nanotube density of states has a number of sharp peaks due to 1-D van Hove singularities, which are absent in graphene and graphite. Despite the presence of these singularities, the overall density of states is similar at high energies, so that the high temperature specific heat should be roughly equal as well. This is to be expected: the high-energy phonons are more reflective of carbon–carbon bonding than the geometry of the graphene sheet.

Thin Films

Thin films are prevalent in the micro and nanoelectronics industry for the fabrication of sensors, actuators and transistors; thus, thermal transport properties affect the performance and reliability of many structures such as transistors, solid-state lasers, sensors, and actuators. Although these devices are traditionally made from bulk crystalline material (silicon), they often contain thin films of oxides, polysilicon, metal, as well as superlattices such as thin-film stacks of GaAs/AlGaAs for lasers.

Single-crystal Thin Films

Silicon-on-insulator (SOI) films with silicon thicknesses of 0.05 μm to 10 μm above a buried silicon dioxide layer are increasingly popular for semiconductor devices due to the increased dielectric isolation associated with SOI/ SOI wafers contain a thin-layer of silicon on an oxide layer and a thin-film of single-crystal silicon, which reduces the effective thermal conductivity of the material by up to 50% as compared to bulk silicon, due to phonon-interface scattering and defects and dislocations in the crystalline structure. Previous studies by Asheghi *et al.*, show a similar trend. Other studies of thin-films show similar thermal effects.

Superlattices

Thermal properties associated with superlattices are critical in the development of semiconductor lasers. Heat conduction of superlattices is less understood than homogeneous thin films. It is theorized that superlattices have a lower thermal conductivity due to impurities from lattice mismatches and at the heterojunctions. Phonon-interface scattering at heterojunctions needs to be considered in this case; fully elastic scattering underestimates the heat conduction, while fully inelastic scattering overestimates the heat conduction. For example, a Si/Ge thin-film superlattice has a greater decrease in thermal conductivity than an AlAs/GaAs film stack due to increased lattice mismatch. A simple estimate of heat conduction of superlattices is:

$$k_n = \left(\frac{C_1 v_1 C_2 v_2}{C_1 v_1 + C_2 v_2}\right)\left(\frac{d_1 + d_2}{2}\right)$$

where C_1 and C_2 are the corresponding heat capacity of film1 and film2 respectively, v_1 and v_2 are the acoustic propagation velocities in film1 and film2, and d_1 and d_2 are the thicknesses of film1 and film2. This model neglects scattering within the layers and assumes fully diffuse, inelastic scattering.

Polycrystalline Films

Polycrystalline films are common in semiconductor devices, as the gate electrode of a field-effect transistor is often made of polycrystalline silicon. If the polysilicon grain sizes are small, internal scattering from grain boundaries can overwhelm the effects of film-boundary scattering. Also, grain boundaries contain more impurities, which result in impurity scattering. Likewise, disordered or amorphous films will experience a severe reduction of thermal conductivity, since the small grain size results in numerous grain-boundary scattering effects. Different deposition methods of amorphous films will result in differences in impurities and grain sizes.

The simplest approach to modeling phonon scattering at grain boundaries is to increase the scattering rate by introducing this equation:

$$\tau_G^{-1} = B\frac{v}{d_G}$$

where B is a dimensionless parameter that correlates with the phonon reflection coefficient at the grain boundaries, d_G is the characteristic grain size, and v is the phonon velocity through the material. A more formal approach to estimating the scattering rate is:

$$\tau_G^{-1} = \frac{2v}{\pi d_G}\left[1 - \exp\left(-\frac{\pi^2}{4}v_G\right)\right]$$

where v_G is the dimensionless grain-boundary scattering strength, defined as:

$$v_G = \sum_j \sigma_j v_j$$

Here σ_j is the cross-section of a grain-boundary area, and v_j is the density of the grain boundary area.

Measuring Thermal Conductivity of Thin Films

There are two approaches to experimentally determine the thermal conductivity of thin films. The goal of experimental metrology of thermal conductivity of thin films is to attain an accurate thermal measurement without disturbing the properties of the thin-film.

Electrical heating is used for thin films which have a lower thermal conductivity than the substrate; it is fairly accurate in measuring out-of-plane conductivity. Often, a resistive heater and thermistor is fabricated on the sample film using a highly conductive metal, such as aluminium. The most straightforward approach would be to apply a steady-state current and measure the change in temperature of adjacent thermistors. A more versatile approach uses an AC signal applied to the electrodes. The third harmonic of the AC signal reveals heating and temperature fluctuations of the material.

Laser heating is a non-contact metrology method, which uses picosecond and nanosecond laser pulses to deliver thermal energy to the substrate. Laser heating uses a pump-probe mechanism; the pump beam introduces energy to the thin-film, as the probe beam picks up the characteristics of how the energy propagates through the film. Laser heating is advantageous because the energy delivered to the film can be precisely controlled; furthermore, the short heating duration decouples the thermal conductivity of the thin film from the substrate.

References

- Holm, Timothy P.; Cheng Chieh Chao; Prinz, Fritz B. (2009). "2009 34th IEEE Photovoltaic Specialists Conference (PVSC)": 000085. ISBN 978-1-4244-2949-3. doi:10.1109/PVSC.2009.5411731

- "Could hemp nanosheets topple graphene for making the ideal supercapacitor?". acs.org. American Chemistry Society. Retrieved 14, August 2020

- M. Corso; Greber, Thomas; Osterwalder, Jürg (2005). "h-BN on Pd(110): a tunable system for selfassembled nanostructures?" Surf Sci 577 (2–3): L78 Bibcode:2005SurSc 577L 78C doi:10 1016/j susc 2005 01 015

- "Simple fabrication method of hierarchical nano-pillars using aluminum anodizing processes". Current Applied Physics. 9: e81–e85. doi:10.1016/j.cap.2008.12.034

- Daniel, Shir; et al. (2006). "Oxidation of silicon nanowires.". Journal of Vacuum Science & Technology (3): 1333–1336

- Liu, M.; Peng, J.; et al. (2016). "Two-dimensional modeling of the self-limiting oxidation in silicon and tungsten nanowires". Theoretical and Applied Mechanics Letters. 6 (5): 195–199. doi:10.1016/j.taml.2016.08.002

- Shao, M.; Duo Duo Ma, D.; Lee, ST (2010). "Silicon nanowires–synthesis, properties, and applications". European Journal of Inorganic Chemistry. 27: 4264–4278

- "Nano-flake structure boosts solar cell performance - Eco-innovation Action Plan - European Commission". Eco-innovation Action Plan. Retrieved 02, February 2020

- R Saito; G Dresselhaus; M S Dresselhaus (22 July 1998). Physical Properties of Carbon Nanotubes. World Scientific. pp. 229–. ISBN 978-1-78326-241-0

- "Synthesis of Single-Crystal Gold Nanosheets of Large Size in Ionic Liquids". The Journal of Physical Chemistry B. 109: 14445–14448. doi:10.1021/jp0520998

- Geim, A. K. (2009). "Graphene: status and prospects". Science. 324 (5934): 1530–1534. PMID 19541989. doi:10.1126/science.1158877

5

Diverse Aspects of Nanotechnology

The mechanism which is used to grow one-dimensional structures like nanowires is vapor-liquid-solid method. This method makes use of foreign element catalytic agent to mediate the growth. This chapter covers fundamental concepts of nanotechnology such as nanopore sequencing, molecular self-assembly, DNA nanotechnology, self-assembled monolayer and supramolecular assembly.

Vapor–Liquid–Solid Method

The vapor–liquid–solid method (VLS) is a mechanism for the growth of one-dimensional structures, such as nanowires, from chemical vapor deposition. The growth of a crystal through direct adsorption of a gas phase on to a solid surface is generally very slow. The VLS mechanism circumvents this by introducing a catalytic liquid alloy phase which can rapidly adsorb a vapor to supersaturation levels, and from which crystal growth can subsequently occur from nucleated seeds at the liquid–solid interface. The physical characteristics of nanowires grown in this manner depend, in a controllable way, upon the size and physical properties of the liquid alloy.

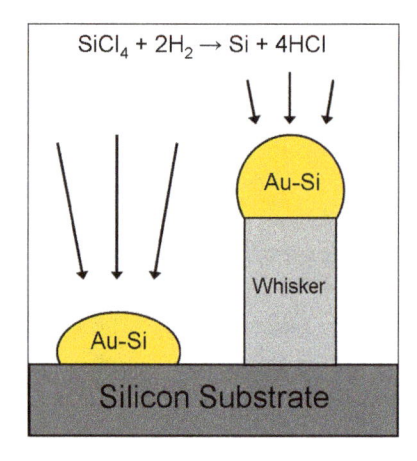

Figure: Schematic illustration of Si whisker growth from the reaction of $SiCl_4$ and H_2 vapor phases. This reaction is catalyzed by gold-silicon droplet deposited on the wafer surface prior to whisker growth.

Historical Background

The VLS mechanism was proposed in 1964 as an explanation for silicon whisker growth from the gas phase in the presence of a liquid gold droplet placed upon a silicon substrate. The explanation was motivated by the absence of axial screw dislocations in the whiskers (which in themselves are a growth mechanism), the requirement of the gold droplet for growth, and the presence of the droplet at the tip of the whisker during the entire growth process.

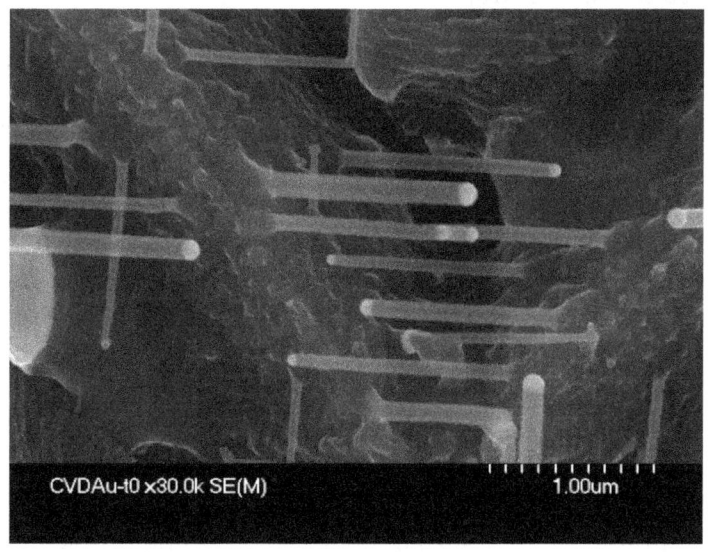

Figure: CVD Growth of Si nanowires using Au particle catalysts.

Introduction

The VLS mechanism is typically described in three stages:

- Preparation of a liquid alloy droplet upon the substrate from which a wire is to be grown.

- Introduction of the substance to be grown as a vapor, which adsorbs on to the liquid surface, and diffuses into the droplet.

- Supersaturation and nucleation at the liquid/solid interface leading to axial crystal growth.

Experimental Technique

The VLS process takes place as follows:

1. A thin (~1–10 nm) Au film is deposited onto a silicon (Si) wafer substrate by sputter deposition or thermal evaporation.

2. The wafer is annealed at temperatures higher than the Au-Si eutectic point, creating Au-Si alloy droplets on the wafer surface (the thicker the Au film, the larger the droplets). Mixing Au with Si greatly reduces the melting temperature of the alloy as compared to the alloy constituents. The melting temperature of

the Au:Si alloy reaches a minimum (~363 °C) when the ratio of its constituents is 4:1 Au:Si, also known as the Au:Si eutectic point.

3. Lithography techniques can also be used to controllably manipulate the diameter and position of the droplets.

4. One-dimensional crystalline nanowires are then grown by a liquid metal-alloy droplet-catalyzed chemical or physical vapor deposition process, which takes place in a vacuum deposition system. Au-Si droplets on the surface of the substrate act to lower the activation energy of normal vapor-solid growth. For example, Si can be deposited by means of a $SiCl_4:H_2$ gaseous mixture reaction (chemical vapor deposition), only at temperatures above 800 °C, in normal vapor-solid growth. Moreover, below this temperature almost no Si is deposited on the growth surface. However, Au particles can form Au-Si eutectic droplets at temperatures above 363 °C and adsorb Si from the vapor state (because Au can form a solid-solution with all Si concentrations up to 100%) until reaching a supersaturated state of Si in Au. Furthermore, nanosized Au-Si droplets have much lower melting points (ref) because the surface area-to-volume ratio is increasing, becoming energetically unfavorable, and nanometer-sized particles act to minimize their surface energy by forming droplets (spheres or half-spheres).

5. Si has a much higher melting point (~1414 °C) than that of the eutectic alloy, therefore Si atoms precipitate out of the supersaturated liquid-alloy droplet at the liquid-alloy/solid-Si interface, and the droplet rises from the surface. This process is illustrated in figure 1.

Typical Features of The VLS Method

- Greatly lowered reaction energy compared to normal vapor-solid growth.

- Wires grow only in the areas activated by the metal catalysts and the size and position of the wires are determined by that of the metal catalysts.

- This growth mechanism can also produce highly anisotropic nanowire arrays from a variety of material.

Requirements for Catalyst Particles

The requirements for catalysts are:

- It must form a liquid solution with the crystalline material to be grown at the nanowire growth temperature.

- The solid solubility of the catalyzing agent is low in the solid and liquid phases of the substrate material.

- The equilibrium vapor pressure of the catalyst over the liquid alloy must be small so that the droplet does not vaporize, shrink in volume (and therefore radius), and decrease the radius of the growing wire until, ultimately, growth is terminated.

- The catalyst must be inert (non-reacting) to the reaction products (during CVD nanowire growth).

- The vapor–solid, vapor–liquid, and liquid–solid interfacial energies play a key role in the shape of the droplets and therefore must be examined before choosing a suitable catalyst; small contact angles between the droplet and solid are more suitable for large area growth, while large contact angles result in the formation of smaller (decreased radius) whiskers.

- The solid-liquid interface must be well-defined crystallographically in order to produce highly directional growth of nanowires. It is also important to point out that the solid-liquid interface cannot, however, be completely smooth. Furthermore, if the solid liquid interface was atomically smooth, atoms near the interface trying to attach to the solid would have no place to attach to until a new island nucleates (atoms attach at step ledges), leading to an extremely slow growth process. Therefore, "rough" solid surfaces, or surfaces containing a large number of surface atomic steps (ideally 1 atom wide, for large growth rates) are needed for deposited atoms to attach and nanowire growth to proceed.

Growth Mechanism

Catalyst Droplet Formation

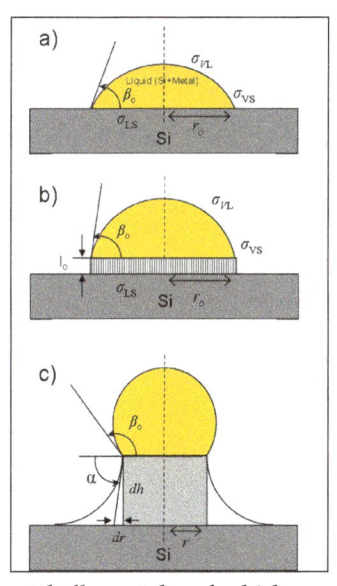

Figure: Schematic illustration of metal-alloy catalyzed whisker growth depicting the catalyst droplet formation during the early stages of whisker growth.

The materials system used, as well as the cleanliness of the vacuum system and therefore the amount of contamination and/or the presence of oxide layers at the droplet and wafer surface during the experiment, both greatly influence the absolute magnitude of the forces present at the droplet/surface interface and, in turn, determine the shape of the droplets. The shape of the droplet, i.e. the contact angle (β_o, see Figure 4) can, be modeled mathematically, however, the actual forces present during growth are extremely difficult to measure experimentally. Nevertheless, the shape of a catalyst particle at the surface of a crystalline substrate is determined by a balance of the forces of surface tension and the liquid–solid interface tension. The radius of the droplet varies with the contact angle as:

$$R = \frac{r_o}{\sin(\beta_o)},$$

where r_o is the radius of the contact area and β_o is defined by a modified Young's equation:

$$\sigma_l \cos(\beta_o) = \sigma_s - \sigma_{ls} - \frac{\tau}{r_o},$$

It is dependent on the surface (σ_s) and liquid–solid interface (σ_{ls}) tensions, as well as an additional line tension (τ) which comes into effect when the initial radius of the droplet is small (nanosized). As a nanowire begins to grow, its height increases by an amount dh and the radius of the contact area decreases by an amount dr (see Figure 4). As the growth continues, the inclination angle at the base of the nanowires (α, set as zero before whisker growth) increases, as does β_o:

$$\sigma_l \cos(\beta_o) = \sigma_s \cos(\alpha) - \sigma_{ls} - \frac{\tau}{r_o}.$$

The line tension therefore greatly influences the catalyst contact area. The most import result from this conclusion is that different line tensions will result in different growth modes. If the line tensions are too large, nanohillock growth will result and thus stop the growth.

Nanowhisker Diameter

The diameter of the nanowire which is grown depends upon the properties of the alloy droplet. The growth of nano-sized wires requires nano-size droplets to be prepared on the substrate. In an equilibrium situation this is not possible as the minimum radius of a metal droplet is given by

$$R_{min} = \frac{2V_l}{RT\ln(s)} \sigma_{lv}$$

where V_l is the molar volume of the droplet, σ_{lv} the liquid-vapor surface energy, and s is the degree of supersaturation of the vapor. This equations restricts the minimum diameter of the droplet, and of any crystals which can be grown from it, under typically conditions to well above the nanometer level. Several techniques to generate smaller droplets have been developed, including the use of monodispersed nanoparticles spread in low dilution on the substrate, and the laser ablation of a substrate-catalyst mixture so to form a plasma which allows well-separated nanoclusters of the catalyst to form as the systems cools.

Whisker Growth Kinetics

During VLS whisker growth, the rate at which whiskers grow is dependent on the whisker diameter: the larger the whisker diameter, the faster the nanowire grows axially. This is because the supersaturation of the metal-alloy catalyst ($Ä\mu$) is the main driving force for nanowhisker growth and decreases with decreasing whisker diameter (also known as the Gibbs-Thomson effect):

$$\Delta\mu = \Delta\mu_o - \frac{4\alpha\Omega}{d}.$$

Again, Δμ is the main driving force for nanowhisker growth (the supersaturation of the metal droplet). More specifically, $\Delta\mu_o$ is the difference between the chemical potential of the depositing species (Si in the above example) in the vapor and solid whisker phase. $\Delta\mu_o$ is the initial difference proceeding whisker growth (when $d \to \infty$), while $\grave{U}$ is the atomic volume of Si and α the specific free energy of the wire surface. Examination of the above equation, indeed reveals that small diameters (<100 nm) exhibit small driving forces for whisker growth while large wire diameters exhibit large driving forces.

Related Growth Techniques

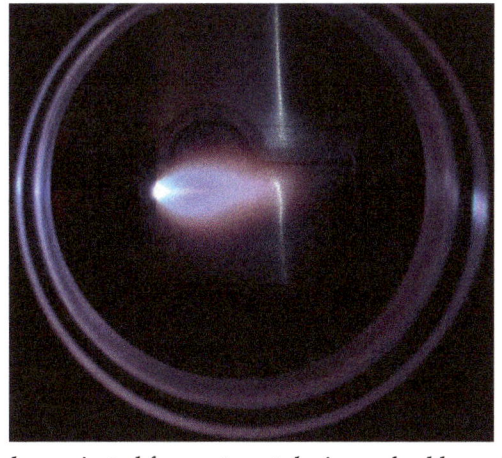

A plasma plume ejected from a target during pulsed laser deposition.

Laser-assisted Growth

Involves the removal of material from metal-containing solid targets by irradiating the surface with high-powered (~100 mJ/pulse) short (10 Hz) laser pulses, usually with wavelengths in the ultraviolet (UV) region of the light spectrum. When such a laser pulse is adsorbed by a solid target, material from the surface region of the target absorbs the laser energy and either (a) evaporates or sublimates from the surface or is (b) converted into a plasma. These particles are easily transferred to the substrate where they can nucleate and grow into nanowires. The laser-assisted growth technique is particularly useful for growing nanowires with high melting temperatures, multicomponent or doped nanowires, as well as nanowires with extremely high crystalline quality. The high intensity of the laser pulse incident at the target allows the deposition of high melting point materials, without having to try to evaporate the material using extremely high temperature resistive or electron bombardment heating. Furthermore, targets can simply be made from a mixture of materials or even a liquid. Finally, the plasma formed during the laser absorption process allows for the deposition of charged particles as well as a catalytic means to lower the activation barrier of reactions between target constituents.

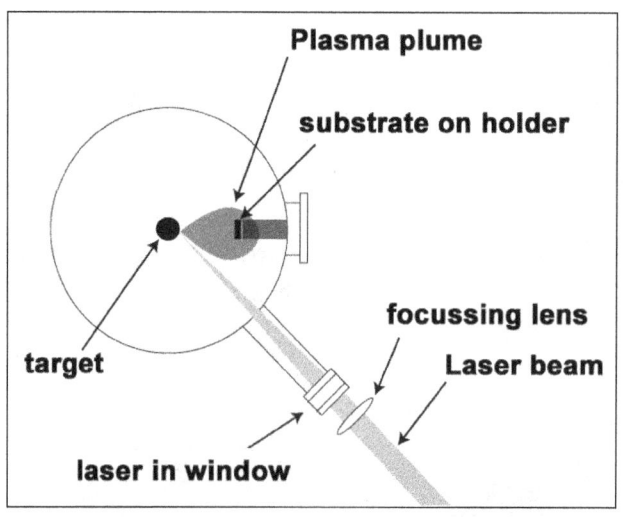

One possible configuration of a PLD deposition chamber.

Thermal Evaporation

Some very interesting nanowires microstructures can be obtained by simply thermally evaporating solid materials. This technique can be carried out in a relatively simple set-up composed of a dual-zone vacuum furnace. The hot end of the furnace contains the evaporating source material, while the evaporated particles are carrier downstream, (by way of a carrier gas) to the colder end of the furnace where they can absorb, nucleate, and grow on a desired substrate.

Metal-catalyzed Molecular Beam Epitaxy

Molecular beam epitaxy (MBE) has been used since 2000 to create high-quality semiconductor wires based on the VLS growth mechanism. However, in metal-catalyzed MBE the metal particles do not catalyze a reaction between precursors but rather adsorb vapor phase particles. This is because the chemical potential of the vapor can be drastically lowered by entering the liquid phase.

MBE is carried out under ultra-high vacuum (UHV) conditions where the mean-free-path (distance between collisions) of source atoms or molecules is on the order of meters. Therefore, evaporated source atoms (from, say, an effusion cell) act as a beam of particles directed towards the substrate. The growth rate of the process is very slow, the deposition conditions are very clean, and as a result four superior capabilities arise, when compared to other deposition methods:

- UHV conditions minimize the amount of oxidation/contamination of the growing structures.

- Relatively low growth temperatures prevent interdiffusion (mixing) of nano-sized heterostructures.

- Very thin-film analysis techniques can be used *in-situ* (during growth), such as reflection high energy electron diffraction (RHEED) to monitor the microstructure at the surface of the substrate as well as the chemical composition, using Auger electron spectroscopy.

Nanopore Sequencing

Nanopore sequencing is a method under development since 1995 for sequencing DNA (determining the order in which nucleotides occur on a strand of DNA).

A nanopore is simply a small hole with an internal diameter of the order of 1 nanometer. Certain porous transmembrane cellular proteins act as nanopores, and nanopores have also been made by etching a somewhat larger hole (several tens of nanometers) in a piece of silicon, and then gradually filling it in using ion-beam sculpting methods which results in a much smaller diameter hole: the nanopore. Graphene is also being explored as a synthetic substrate for solid-state nanopores.

The theory behind nanopore sequencing is that when a nanopore is immersed in a conducting fluid and a potential (voltage) is applied across it, an electric current due to conduction of ions through the nanopore can be observed. The amount of current is very sensitive to the size and shape of the nanopore. If single nucleotides (bases), strands of DNA or other molecules pass through or near the nanopore, this can create a characteristic change in the magnitude of the current through the nanopore.

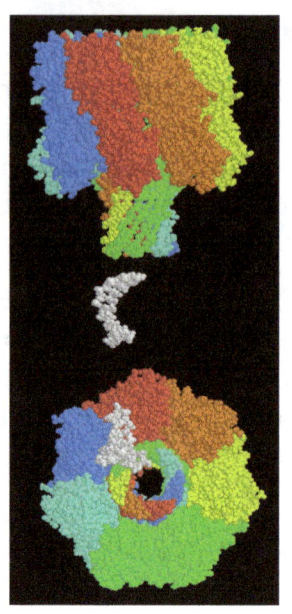

Alpha-hemolysin pore (made up of 7 identical subunits in 7 colors) and 12-mer single-stranded DNA (in white) on the same scale to illustrate DNA effects on conductance when moving through a nanopore. Below is an orthogonal view of the same molecules.

Background

DNA could be passed through the nanopore for various reasons. For example, electrophoresis might attract the DNA towards the nanopore, and it might eventually pass through it. Or, enzymes attached to the nanopore might guide DNA towards the nanopore. The scale of the nanopore means that the DNA may be forced through the hole as a long string, one base at a time, rather like threading through the eye of a needle. As it does so, each nucleotide on the DNA molecule may obstruct the nanopore to a different, characteristic degree. The amount of current which can pass through the nanopore at any given moment therefore varies depending on whether the nanopore is blocked by an A, a C, a G or a T or a section of DNA that includes more than one of these bases (kmer). The change in the current through the nanopore as the DNA molecule passes through the nanopore represents a direct reading of the DNA sequence. This is the 'strand sequencing' approach commercialised by Oxford Nanopore Technologies.

Alternatively, a nanopore might be used to identify individual DNA bases as they pass through the nanopore in the correct order - this approach was published but not commercially developed by Oxford Nanopore Technologies and Professor Hagan Bayley.

Using Nanopore sequencing, a single molecule of DNA can be sequenced directly using a nanopore, without the need for an intervening PCR amplification step or a chemical labelling step or the need for optical instrumentation to identify the chemical label. The versatility of the nanopore concept is underlined by the fact that it has also been proposed for detection of life on other planets, since it is not necessarily restricted to the detection of the genetic information carrier DNA, but in general can be applied to sequence chain-like genetic information carriers without knowing the exact structure of their building blocks. Nanopore-based DNA analysis techniques are being industrially developed by Oxford Nanopore Technologies (developing strand sequencing using protein nanopores, and solid-state sequencing through internal R&D and collaborations with academic institutions). The MinION was made commercially available in June 2015 after having been introduced in 2014 in an early access programme. Publications on the method outline its use in rapid identification of viral pathogens, monitoring ebola, environmental monitoring, food safety monitoring, monitoring of antibiotic resistance, haplotyping and other applications. Other companies have noted that they have Nanopore programmes, NabSys (using a library of DNA probes and using nanopores to detect where these probes have hybridized to single stranded DNA) and NobleGen (using nanopores in combination with fluorescent labels). IBM has noted research projects on computer simulations of translocation of a DNA strand through a solid-state nanopore, but not projects on identifying the DNA bases on that strand.

Types of Nanopores

Alpha Hemolysin

Alpha hemolysin (αHL), a nanopore from bacteria that causes lysis of red blood cells, has been studied for over 15 years. To this point, studies have shown that all four bases can be identified using ionic current measured across the αHL pore. The structure of αHL is advantageous to identify specific bases moving through the pore. The αHL pore is ~10 nm long, with two distinct 5 nm sections. The upper section consists of a larger, vestibule-like structure and the lower section consists of three possible recognition sites (R1, R2, R3), and is able to discriminate between each base.

Sequencing using αHL has been developed through basic study and structural mutations, moving towards the sequencing of very long reads. Protein mutation of αHL has improved the detection abilities of the pore. The next proposed step is to bind an exonuclease onto the αHL pore. The enzyme would periodically cleave single bases, enabling the pore to identify successive bases. Coupling an exonuclease to the biological pore would slow the translocation of the DNA through the pore, and increase the accuracy of data acquisition.

Notably, theorists have shown that sequencing via exonuclease enzymes as described here is not feasible. This is mainly due to diffusion related effects imposing a limit on the capture probability of each nucleotide as it is cleaved. This results in a significant probability that a nucleotide is either not captured before it diffuses into the bulk or captured out of order, and therefore is not properly sequenced by the nanopore, leading to insertion and deletion errors. Therefore, major changes are needed to this method before it can be considered a viable strategy.

A recent study has pointed to the ability of αHL to detect nucleotides at two separate sites in the lower half of the pore. The R1 and R2 sites enable each base to be monitored twice as it moves through the pore, creating 16 different measurable ionic current values instead of 4. This method improves upon the single read through the nanopore by doubling the sites that the sequence is read per nanopore.

MspA

Mycobacterium smegmatis porin A (MspA) is the second biological nanopore currently being investigated for DNA sequencing. The MspA pore has been identified as a potential improvement over αHL due to a more favorable structure. The pore is described as a goblet with a thick rim and a diameter of 1.2 nm at the bottom of the pore. A natural MspA, while favorable for DNA sequencing because of shape and diameter, has a negative core that prohibited single stranded DNA(ssDNA) translocation. The natural nanopore was modified to improve translocation by replacing three negatively charged aspartic acids with neutral asparagines.

The electric current detection of nucleotides across the membrane has been shown to be tenfold more specific than αHL for identifying bases. Utilizing this improved specificity, a group at the University of Washington has proposed using double stranded DNA (dsDNA) between each single stranded molecule to hold the base in the reading section of the pore. The dsDNA would halt the base in the correct section of the pore and enable identification of the nucleotide. A recent grant has been awarded to a collaboration from UC Santa Cruz, the University of Washington, and Northeastern University to improve the base recognition of MspA using phi29 polymerase in conjunction with the pore.

CsgG

In March 2016, the lab of Han Remaut (VIB/Vrije Universiteit Brussels) announced a collaboration with Oxford Nanopore Technologies using the protein nanopore CsgG for DNA sequencing. Oxford Nanopore simultaneously announced the release of this new chemistry for its MinION device. – designed to improve accuracy and yields. The R9 nanopore was launched in late May 2016 and early reports indicate higher performance levels. More information can be found at the Oxford Nanopore page.

Fluorescence

An effective technique to determine a DNA sequence has been developed using solid state nanopores and fluorescence. This fluorescence sequencing method converts each base into a characteristic representation of multiple nucleotides which bind to a fluorescent probe strand-forming dsDNA. With the two color system proposed, each base is identified by two separate fluorescences, and will therefore be converted into two specific sequences. Probes consist of a fluorophore and quencher at the start and end of each sequence, respectively. Each fluorophore will be extinguished by the quencher at the end of the preceding sequence. When the dsDNA is translocating through a solid state nanopore, the probe strand will be stripped off, and the upstream fluorophore will fluoresce.

This sequencing method has a capacity of 50-250 bases per second per pore, and a four color fluorophore system (each base could be converted to one sequence instead of two), will sequence over 500 bases per second. Advantages of this method are based on the clear sequencing readout—using a camera instead of noisy current methods. However, the method does require sample preparation to convert each base into an expanded binary code before sequencing. Instead of one base being identified as it translocates through the pore, ~12 bases are required to find the sequence of one base.

Electron Tunneling

Measurement of electron tunneling through bases as ssDNA translocates through the nanopore is an improved solid state nanopore sequencing method. Most research has focused on proving bases could be determined using electron tunneling. These studies were conducted using a scanning probe microscope as the sensing electrode, and have proved that bases can be identified by specific tunneling currents. After the proof of principle research, a functional system must be created to couple the solid state pore and sensing devices.

Researchers at the Harvard Nanopore group have engineered solid state pores with single walled carbon nanotubes across the diameter of the pore. Arrays of pores are created and chemical vapor deposition is used to create nanotubes that grow across the array. Once a nanotube has grown across a pore, the diameter of the pore is adjusted to the desired size. Successful creation of a nanotube coupled with a pore is an important step towards identifying bases as the ssDNA translocates through the solid state pore.

Another method is the use of nanoelectrodes on either side of a pore. The electrodes are specifically created to enable a solid state nanopore's formation between the two electrodes. This technology could be used to not only sense the bases but to help control base translocation speed and orientation.

Challenges

One challenge for the 'strand sequencing' method was in refining the method to improve its resolution to be able to detect single bases. In the early papers methods, a nucleotide needed to be repeated in a sequence about 100 times successively in order to produce a measurable characteristic change. This low resolution is because the DNA strand moves rapidly at the rate of 1 to 5μs per base through the nanopore. This makes recording difficult and prone to background noise, failing in obtaining single-nucleotide resolution. The problem is being tackled by either improving the recording technology or by controlling the speed of DNA strand by various protein engineering strategies and

Oxford Nanopore employs a 'kmer approach', analyzing more than one base at any one time so that stretches of DNA are subject to repeat interrogation as the strand moves through the nanopore one base at a time. Various techniques including algorithmic have been used to improve the performance of the MinION technology since it was first made available to users. More recently effects of single bases due to secondary structure or released mononucleotides have been shown. Professor Hagan Bayley, founder of Oxford Nanopore, proposed in 2010 that creating two recognition sites within an alpha hemolysin pore may confer advantages in base recognition.

One challenge for the 'exonuclease approach', initially developed but not progressed by Oxford Nanopore Technologies, where a processive enzyme feeds individual bases, in the correct order, into the nanopore, is to integrate the exonuclease and the nanopore detection systems. In particular, the problem is that when an exonuclease hydrolyzes the phosphodiester bonds between nucleotides in DNA, the subsequently released nucleotide is not necessarily guaranteed to directly move in to, say, a nearby alpha-hemolysin nanopore. One idea is to attach the exonuclease to the nanopore, perhaps through biotinylation to the beta barrel hemolsyin. The central pore of the protein may be lined with charged residues arranged so that the positive and negative charges appear on opposite sides of the pore. However, this mechanism is primarily discriminatory and does not constitute a mechanism to guide nucleotides down some particular path.

Commercialization

Agilent Laboratories was the first to license and develop nanopores but does not have any current disclosed research in the area.

The company Oxford Nanopore Technologies now sells the MinION portable DNA sequencing device and is introducing the PromethION high-throughput device into an early access programme in 2016. In 2016 it announced that the first mobile-phone sequencer, SmidgION, was in development. In 2008 licensed technology from Harvard, UCSC and other universities and is developing protein and solid state nanopore technology with the aim of analysing a range of biological molecules including DNA, RNA and proteins. They revealed the idea of their first working device in February 2012. The Company introduced its pocket-sized sensing device MinION to an early-access community in early 2014 and made it commercially available in May 2015. A variety of applications for the device have now been developed including environmental or pathogen monitoring, genome research, investigation of fetal DNA.

Sequenom licensed nanopore technology from Harvard in 2007 using an approach that combines nanopores and fluorescent labels. This technology was subsequently licensed to Noblegen.

NABsys was spun out of Brown University and is researching nanopores as a method of identifying areas of single stranded DNA that have been hybridized with specific DNA probes.

Genia applied for a patent on a method of nanopore sequencing which uses a series of RNA "speed-bump" molecules to infer the sequence of the molecule based on the pattern of delayed progression.

Molecular Self-assembly

Molecular self-assembly is the process by which molecules adopt a defined arrangement without guidance or management from an outside source. There are two types of self-assembly. These are intramolecular self-assembly and intermolecular self-assembly. Commonly, the term molecular self-assembly refers to intermolecular self-assembly, while the intramolecular analog is more commonly called folding.

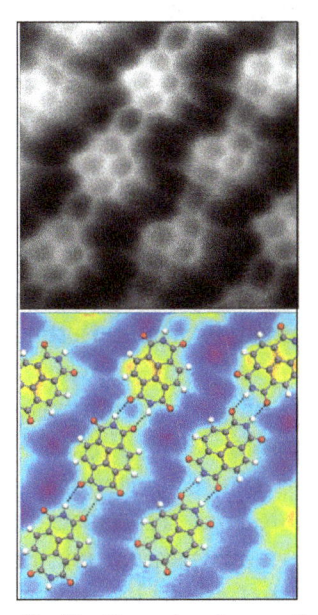

AFM image of napthalenetetracarboxylic diimide molecules on silver interacting via hydrogen bonding (77 K).

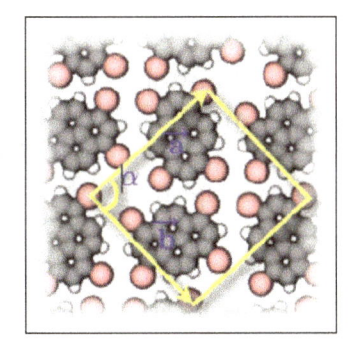

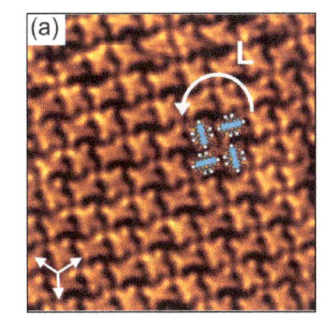

STM image of self-assembled Br_4-pyrene molecules on Au(111) surface (top) and its model (bottom; pink spheres are Br atoms).

Supramolecular Systems

Molecular self-assembly is a key concept in supramolecular chemistry. This is because assembly of molecules in such systems is directed through noncovalent interactions (e.g., hydrogen bonding, metal coordination, hydrophobic forces, van der Waals forces, π-π interactions, and/or electrostatic) as well as electromagnetic interactions. Common examples include the formation of micelles, vesicles, liquid crystal phases, and Langmuir monolayers by surfactant molecules. Further examples of supramolecular assemblies demonstrate that a variety of different shapes and sizes can be obtained using molecular self-assembly.

Molecular self-assembly allows the construction of challenging molecular topologies. One example is Borromean rings, interlocking rings wherein removal of one ring unlocks each of the other rings. DNA has been used to prepare a molecular analog of Borromean rings. More recently, a similar structure has been prepared using non-biological building blocks. While a mechanistic understanding of how supramolecular self-assembly occurs remains largely unknown, both experimental and theoretic work has been pursued on this topic.

Biological Systems

Molecular self-assembly underlies the construction of biologic macromolecular assemblies in living organisms, and so is crucial to the function of cells. It is exhibited in the self-assembly of lipids to form the membrane, the formation of double helical DNA through hydrogen bonding of the individual strands, and the assembly of proteins to form quaternary structures. Molecular self-assembly of incorrectly folded proteins into insoluble amyloid fibers is responsible for infectious prion-related neurodegenerative diseases. Molecular self-assembly of nanoscale structures plays a role in the growth of the remarkable β-keratin lamellae/setae/spatulae structures used to give geckos the ability to climb walls and adhere to ceilings and rock overhangs.

Nanotechnology

Molecular self-assembly is an important aspect of bottom-up approaches to nanotechnology. Using molecular self-assembly the final (desired) structure is programmed in the shape and functional groups of the molecules. Self-assembly is referred to as a 'bottom-up' manufacturing technique in contrast to a 'top-down' technique such as lithography where the desired final structure is carved from a larger block of matter. In the speculative vision of molecular nanotechnology, microchips of the future might be made by molecular self-assembly. An advantage to constructing nanostructure using molecular self-assembly for biological materials is that they will degrade back into individual molecules that can be broken down by the body.

DNA Nanotechnology

DNA nanotechnology is an area of current research that uses the bottom-up, self-assembly approach for nanotechnological goals. DNA nanotechnology uses the unique molecular recognition properties of DNA and other nucleic acids to create self-assembling branched DNA complexes with useful properties. DNA is thus used as a structural material rather than as a carrier of biological information, to make structures such as two-dimensional periodic lattices (both tile-based as well as using the "DNA origami" method) and three-dimensional structures in the shapes of polyhedra. These DNA structures have also been used as templates in the assembly of other molecules such as gold nanoparticles and streptavidin proteins.

Two-dimensional Monolayers

The spontaneous assembly of a single layer of molecules at interfaces is usually referred to as two-dimensional self-assembly. Early direct proofs showing that molecules can assembly into higher-order architectures at solid interfaces came with the development of scanning tunneling microscopy and shortly thereafter. Eventually two strategies became popular for the self-assembly of 2D architectures, namely self-assembly following ultra-high-vacuum deposition and annealing and self-assembly at the solid-liquid interface. The design of molecules and conditions leading to the formation of highly-crystalline architectures is considered today a form of 2D crystal engineering at the nanoscopic scale.

DNA Nanotechnology

DNA nanotechnology is the design and manufacture of artificial nucleic acid structures for technological uses. In this field, nucleic acids are used as non-biological engineering materials for nanotechnology rather than as the carriers of genetic information in living cells. Researchers in the field have created static structures such as two- and three-dimensional crystal lattices, nanotubes, polyhedra, and arbitrary shapes, and functional devices such as molecular machines and DNA computers. The field is beginning to be used as a tool to solve basic science problems in structural biology and biophysics, including applications in X-ray crystallography and nuclear magnetic resonance spectroscopy of proteins to determine structures. Potential applications in molecular scale electronics and nanomedicine are also being investigated.

The conceptual foundation for DNA nanotechnology was first laid out by Nadrian Seeman in the early 1980s, and the field began to attract widespread interest in the mid-2000s. This use of nucleic acids is enabled by their strict base pairing rules, which cause only portions of strands with complementary base sequences to bind together to form strong, rigid double helix structures. This allows for the rational design of base sequences that will selectively assemble to form complex target structures with precisely controlled nanoscale features. Several assembly methods are used to make these structures, including tile-based structures that assemble from smaller structures, folding structures using the DNA origami method, and dynamically reconfigurable structures using strand displacement methods. While the field's name specifically references DNA, the same principles have been used with other types of nucleic acids as well, leading to the occasional use of the alternative name *nucleic acid nanotechnology*.

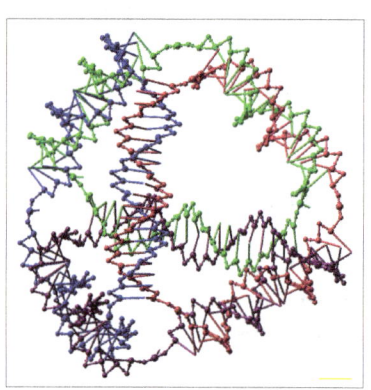

DNA nanotechnology involves forming artificial, designed nanostructures out of nucleic acids, such as this DNA tetrahedron. Each edge of the tetrahedron is a 20 base pair DNA double helix, and each vertex is a three-arm junction. The 4 DNA strands that form the 4 tetrahedral faces are color-coded.

Fundamental Concepts

These four strands associate into a DNA four-arm junction because this structure maximizes the number of correct base pairs, with A matched to T and C matched to G. See this image for a more realistic model of the four-arm junction showing its tertiary structure.

This double-crossover (DX) supramolecular complex consists of five DNA single strands that form two double-helical domains, on the top and the bottom in this image. There are two crossover points where the strands cross from one domain into the other.

Properties of Nucleic Acids

Nanotechnology is often defined as the study of materials and devices with features on a scale below 100 nanometers. DNA nanotechnology, specifically, is an example of bottom-up molecular self-assembly, in which molecular components spontaneously organize into stable structures; the particular form of these structures is induced by the physical and chemical properties of the components selected by the designers. In DNA nanotechnology, the component materials are strands of nucleic acids such as DNA; these strands are often synthetic and are almost always used outside the context of a living cell. DNA is well-suited to nanoscale construction because the binding between two nucleic acid strands depends on simple base pairing rules which are well understood, and form the specific nanoscale structure of the nucleic acid double helix. These qualities make the assembly of nucleic acid structures easy to control through nucleic acid design. This property is absent in other materials used in nanotechnology, including proteins, for which protein design is very difficult, and nanoparticles, which lack the capability for specific assembly on their own.

The structure of a nucleic acid molecule consists of a sequence of nucleotides distinguished by which nucleobase they contain. In DNA, the four bases present are adenine (A), cytosine (C), guanine (G), and thymine (T). Nucleic acids have the property that two molecules will only bind to each other to form a double helix if the two sequences are complementary, meaning that they form matching sequences of base pairs, with A only binding to T, and C only to G. Because the formation of correctly matched base pairs is energetically favorable, nucleic acid strands are expected in most cases to bind to each other in the conformation that maximizes the number of correctly paired bases. The sequences of bases in a system of strands thus determine the pattern of binding and the overall structure in an easily controllable way. In DNA nanotechnology, the base sequences of strands are rationally designed by researchers so that the base pairing interactions cause the strands to assemble in the desired conformation. While DNA is the dominant material used, structures incorporating other nucleic acids such as RNA and peptide nucleic acid (PNA) have also been constructed.

Subfields

DNA nanotechnology is sometimes divided into two overlapping subfields: structural DNA nanotechnology and dynamic DNA nanotechnology. Structural DNA nanotechnology, sometimes abbreviated as SDN, focuses on synthesizing and characterizing nucleic acid complexes and materials that assemble into a static, equilibrium end state. On the other hand, dynamic DNA nanotechnology focuses on complexes with useful non-equilibrium behavior such as the ability to reconfigure based on a chemical or physical stimulus. Some complexes, such as nucleic acid nanomechanical devices, combine features of both the structural and dynamic subfields.

The complexes constructed in structural DNA nanotechnology use topologically branched nucleic acid structures containing junctions. (In contrast, most biological DNA exists as an unbranched double helix.) One of the simplest branched structures is a four-arm junction that consists of four individual DNA strands, portions of which are complementary in a specific pattern. Unlike in natural Holliday junctions, each arm in the artificial immobile four-arm junction has a different base sequence, causing the junction point to be fixed at a certain position. Multiple junctions can be combined in the same complex, such as in the widely used double-crossover (DX) structural motif, which contains two parallel double helical domains with individual strands crossing between the domains at two crossover points. Each crossover point is, topologically, a four-arm junction, but is constrained to one orientation, in contrast to the flexible single four-arm junction, providing a rigidity that makes the DX motif suitable as a structural building block for larger DNA complexes.

Dynamic DNA nanotechnology uses a mechanism called toehold-mediated strand displacement to allow the nucleic acid complexes to reconfigure in response to the addition of a new nucleic acid strand. In this reaction, the incoming strand binds to a single-stranded toehold region of a double-stranded complex, and then displaces one of the strands bound in the original complex through a branch migration process. The overall effect is that one of the strands in the complex is replaced with another one. In addition, reconfigurable structures and devices can be made using functional nucleic acids such as deoxyribozymes and ribozymes, which can perform chemical reactions, and aptamers, which can bind to specific proteins or small molecules.

Structural DNA Nanotechnology

Structural DNA nanotechnology, sometimes abbreviated as SDN, focuses on synthesizing and characterizing nucleic acid complexes and materials where the assembly has a static, equilibrium endpoint. The nucleic acid double helix has a robust, defined three-dimensional geometry that makes it possible to predict and design the structures of more complicated nucleic acid complexes. Many such structures have been created, including two- and three-dimensional structures, and periodic, aperiodic, and discrete structures.

Extended Lattices

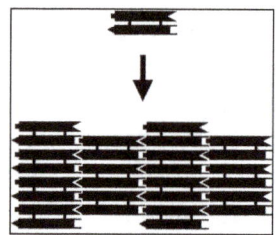

The assembly of a DX array. *Left*, schematic diagram. Each bar represents a double-helical domain of DNA, with the shapes representing complementary sticky ends. The DX complex at top will combine with other DX complexes into the two-dimensional array shown at bottom. *Right*, an atomic force microscopy image of the assembled array. The individual DX tiles are clearly visible within the assembled structure. The field is 150 nm across.

Small nucleic acid complexes can be equipped with sticky ends and combined into larger two-dimensional periodic lattices containing a specific tessellated pattern of the individual molecular tiles. The earliest example of this used double-crossover (DX) complexes as the basic tiles, each containing four sticky ends designed with sequences that caused the DX units to combine into periodic two-dimensional flat sheets that are essentially rigid two-dimensional crystals of DNA. Two-dimensional arrays have been made from other motifs as well, including the Holliday junction rhombus lattice, and various DX-based arrays making use of a double-cohesion scheme. The top two images at right show examples of tile-based periodic lattices.

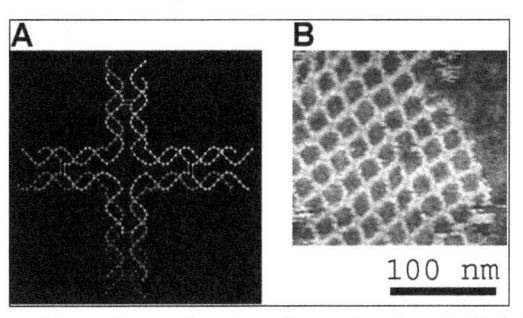

Left, a model of a DNA tile used to make another two-dimensional periodic lattice. *Right*, an atomic force micrograph of the assembled lattice.

Two-dimensional arrays can be made to exhibit aperiodic structures whose assembly implements a specific algorithm, exhibiting one form of DNA computing. The DX tiles can have their sticky end sequences chosen so that they act as Wang tiles, allowing them to perform computation. A DX array whose assembly encodes an XOR operation has been demonstrated; this allows the DNA array to implement a cellular automaton that generates a fractal known as the Sierpinski gasket. The third image at right shows this type of array. Another system has the function of a binary counter, displaying a representation of increasing binary numbers as it grows. These results show that computation can be incorporated into the assembly of DNA arrays.

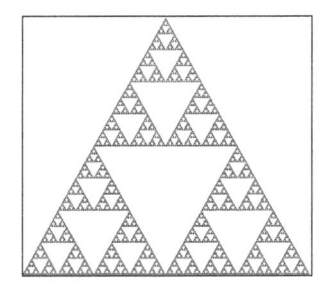

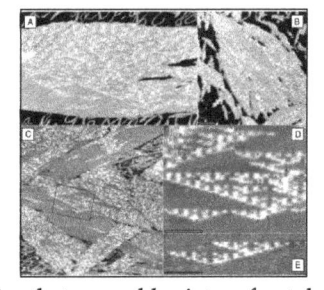

An example of an aperiodic two-dimensional lattice that assembles into a fractal pattern. *Left*, the Sierpinski gasket fractal. *Right*, DNA arrays that display a representation of the Sierpinski gasket on their surfaces.

DX arrays have been made to form hollow nanotubes 4–20 nm in diameter, essentially two-dimensional lattices which curve back upon themselves. These DNA nanotubes are somewhat similar in size and shape to carbon nanotubes, and while they lack the electrical conductance of carbon nanotubes, DNA nanotubes are more easily modified and connected to other structures. One of many schemes for constructing DNA nanotubes uses a lattice of curved DX tiles that curls around itself and closes into a tube. In an alternative method that allows the circumference to be specified in a simple, modular fashion using single-stranded tiles, the rigidity of the tube is an emergent property.

Forming three-dimensional lattices of DNA was the earliest goal of DNA nanotechnology, but this proved to be one of the most difficult to realize. Success using a motif based on the concept of tensegrity, a balance between tension and compression forces, was finally reported in 2009.

Discrete Structures

Researchers have synthesized many three-dimensional DNA complexes that each have the connectivity of a polyhedron, such as a cube or octahedron, meaning that the DNA duplexes trace the edges of a polyhedron with a DNA junction at each vertex. The earliest demonstrations of DNA polyhedra were very work-intensive, requiring multiple ligations and solid-phase synthesis steps to create catenated polyhedra. Subsequent work yielded polyhedra whose synthesis was much easier. These include a DNA octahedron made from a long single strand designed to fold into the correct conformation, and a tetrahedron that can be produced from four DNA strands in one step, pictured at the top of this article.

Nanostructures of arbitrary, non-regular shapes are usually made using the DNA origami method. These structures consist of a long, natural virus strand as a "scaffold", which is made to fold into the desired shape by computationally designed short "staple" strands. This method has the advantages of being easy to design, as the base sequence is predetermined by the scaffold strand sequence, and not requiring high strand purity and accurate stoichiometry, as most other DNA nanotechnology methods do. DNA origami was first demonstrated for two-dimensional shapes, such as a smiley face and a coarse map of the Western Hemisphere. Solid three-dimensional structures can be made by using parallel DNA helices arranged in a honeycomb pattern, and structures with two-dimensional faces can be made to fold into a hollow overall three-dimensional shape, akin to a cardboard box. These can be programmed to open and reveal or release a molecular cargo in response to a stimulus, making them potentially useful as programmable molecular cages.

Templated Assembly

Nucleic acid structures can be made to incorporate molecules other than nucleic acids, sometimes called heteroelements, including proteins, metallic nanoparticles, quantum dots, and fullerenes. This allows the construction of materials and devices with a range of functionalities much greater than is possible with nucleic acids alone. The goal is to use the self-assembly of the nucleic acid structures to template the assembly of the nanoparticles hosted on them, controlling their position and in some cases orientation. Many of these schemes use a covalent attachment scheme, using oligonucleotides with amide or thiol functional groups as a chemical handle to bind the heteroelements. This covalent binding scheme has been used to arrange gold nanoparticles on a DX-based array, and to arrange streptavidin protein molecules into specific patterns on a DX array. A non-covalent hosting scheme using Dervan polyamides on a DX array was used to arrange streptavidin proteins in a specific pattern on a DX array. Carbon nanotubes have been hosted on DNA arrays in a pattern allowing the assembly to act as a molecular electronic device, a carbon nanotube field-effect transistor. In addition, there are nucleic acid metallization methods, in which the nucleic acid is replaced by a metal which assumes the general shape of the original nucleic acid structure, and schemes for using nucleic acid nanostructures as lithography masks, transferring their pattern into a solid surface.

Dynamic DNA Nanotechnology

Dynamic DNA nanotechnology focuses on forming nucleic acid systems with designed dynamic functionalities related to their overall structures, such as computation and mechanical motion. There is some overlap between structural and dynamic DNA nanotechnology, as structures can be formed through annealing and then reconfigured dynamically, or can be made to form dynamically in the first place.

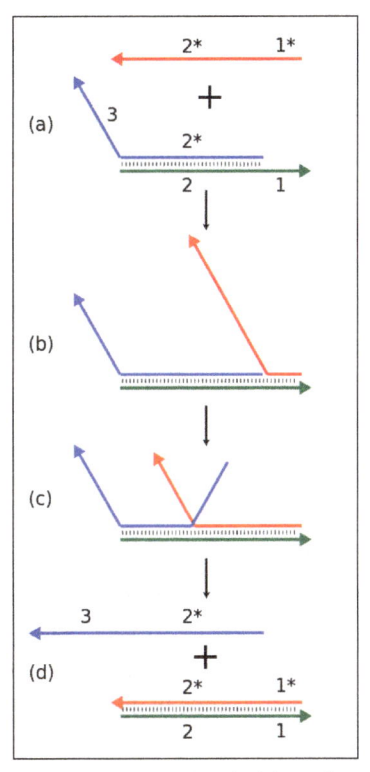

Dynamic DNA nanotechnology often makes use of toehold-mediated strand displacement reactions. In this example, the red strand binds to the single stranded toehold region on the green strand (region 1), and then in a branch migration process across region 2, the blue strand is displaced and freed from the complex. Reactions like these are used to dynamically reconfigure or assemble nucleic acid nanostructures. In addition, the red and blue strands can be used as signals in a molecular logic gate.

Nanomechanical Devices

DNA complexes have been made that change their conformation upon some stimulus, making them one form of nanorobotics. These structures are initially formed in the same way as the static structures made in structural DNA nanotechnology, but are designed so that dynamic reconfiguration is possible after the initial assembly. The earliest such device made use of the transition between the B-DNA and Z-DNA forms to respond to a change in buffer conditions by undergoing a twisting motion. This reliance on buffer conditions, however, caused all devices to change state at the same time. Subsequent systems could change states based upon the presence of control strands, allowing multiple devices to be independently operated in solution. Some examples of such systems are a "molecular tweezers" design that has an open and a closed state, a device that could switch from a paranemic-crossover (PX) conformation to a double-junction (JX2) conformation, undergoing rotational motion in the process, and a two-dimensional array that could dynamically expand and contract in response to control strands. Structures have also been made that dynamically open or close, potentially acting as a molecular cage to release or reveal a functional cargo upon opening.

DNA walkers are a class of nucleic acid nanomachines that exhibit directional motion along a linear track. A large number of schemes have been demonstrated. One strategy is to control the motion of the walker along the track using control strands that need to be manually added in sequence. Another approach is to make use of restriction enzymes or deoxyribozymes to cleave the strands and cause the walker to move forward, which has the advantage of running autonomously. A later system could walk upon a two-dimensional surface rather than a linear track, and demonstrated the ability to selectively pick up and move molecular cargo. Additionally, a linear walker has been demonstrated that performs DNA-templated synthesis as the walker advances along the track, allowing autonomous multistep chemical synthesis directed by the walker. The synthetic DNA walkers' function is similar to that of the proteins dynein and kinesin.

Strand Displacement Cascades

Cascades of strand displacement reactions can be used for either computational or structural purposes. An individual strand displacement reaction involves revealing a new sequence in response to the presence of some initiator strand. Many such reactions can be linked into a cascade where the newly revealed output sequence of one reaction can initiate another strand displacement reaction elsewhere. This in turn allows for the construction of chemical reaction networks with many components, exhibiting complex computational and information processing abilities. These cascades are made energetically favorable through the formation of new base pairs, and the entropy gain from disassembly reactions. Strand displacement cascades allow isothermal operation of the assembly or computational process, in contrast to traditional nucleic acid assembly's requirement for a thermal annealing step, where the temperature is raised and then slowly lowered to ensure proper formation of the desired structure. They can also support catalytic function of the initiator species, where less than one equivalent of the initiator can cause the reaction to go to completion.

Strand displacement complexes can be used to make molecular logic gates capable of complex computation. Unlike traditional electronic computers, which use electric current as inputs and outputs, molecular computers use the concentrations of specific chemical species as signals. In the case of nucleic acid strand displacement circuits, the signal is the presence of nucleic acid strands that are released or consumed by binding and unbinding events to other strands in displacement complexes. This approach has been used to make logic gates such as AND, OR, and NOT gates. More recently, a four-bit circuit was demonstrated that can compute the square root of the integers 0–15, using a system of gates containing 130 DNA strands.

Another use of strand displacement cascades is to make dynamically assembled structures. These use a hairpin structure for the reactants, so that when the input strand binds, the newly revealed sequence is on the same molecule rather than disassembling. This allows new opened hairpins to be added to a growing complex. This approach has been used to make simple structures such as three- and four-arm junctions and dendrimers.

Applications

DNA nanotechnology provides one of the few ways to form designed, complex structures with precise control over nanoscale features. The field is beginning to see application to solve basic science problems in structural biology and biophysics. The earliest such application envisaged for the field, and one still in development, is in crystallography, where molecules that are difficult to crystallize in isolation could be arranged within a three-dimensional nucleic acid lattice, allowing determination of their structure. Another application is the use of DNA origami rods to replace liquid crystals in residual dipolar coupling experiments in protein NMR spectroscopy; using DNA origami is advantageous because, unlike liquid crystals, they are tolerant of the detergents needed to suspend membrane proteins in solution. DNA walkers have been used as nanoscale assembly lines to move nanoparticles and direct chemical synthesis. Further, DNA origami structures have aided in the biophysical studies of enzyme function and protein folding.

DNA nanotechnology is moving toward potential real-world applications. The ability of nucleic acid arrays to arrange other molecules indicates its potential applications in molecular scale electronics. The assembly of a nucleic acid structure could be used to template the assembly of a molecular electronic elements such as molecular wires, providing a method for nanometer-scale control of the placement and overall architecture of the device analogous to a molecular breadboard. DNA nanotechnology has been compared to the concept of programmable matter because of the coupling of computation to its material properties.

In a study conducted by a group of scientists from iNANO center and CDNA Center in Aarhus university (Aarhus), researchers were able to construct a small multi-switchable 3D DNA Box Origami. The proposed nanoparticle was characterized by atomic force microscopy (AFM), transmission electron microscopy (TEM) and Förster resonance energy transfer (FRET). The constructed box was shown to have a unique reclosing mechanism, which enabled it to repeatedly open and close in response to a unique set of DNA or RNA keys. The authors proposed that this "DNA device can potentially be used for a broad range of applications such as controlling the function of single molecules, controlled drug delivery, and molecular computing.".

There are potential applications for DNA nanotechnology in nanomedicine, making use of its ability to perform computation in a biocompatible format to make "smart drugs" for targeted drug delivery. One such system being investigated uses a hollow DNA box containing proteins that induce apoptosis, or cell death, that will only open when in proximity to a cancer cell. There has additionally been interest in expressing these artificial structures in engineered living bacterial cells, most likely using the transcribed RNA for the assembly, although it is unknown whether these complex structures are able to efficiently fold or assemble in the cell's cytoplasm. If successful, this could enable directed evolution of nucleic acid nanostructures. Scientists at Oxford University reported the self-assembly of four short strands of synthetic DNA into a cage which can enter cells and survive for at least 48 hours. The fluorescently labeled DNA tetrahedra were found to remain intact in the laboratory cultured human kidney cells despite the attack by cellular enzymes after two days. This experiment showed the potential of drug delivery inside the living cells using the DNA 'cage'. A DNA tetrahedron was used to deliver RNA Interference (RNAi) in a mouse model, reported a team of researchers in MIT. Delivery of the interfering RNA for treatment has showed some success using polymer or lipid, but there are limits of safety and imprecise targeting, in addition to short shelf life in the blood stream. The DNA nanostructure created by the team consists of six strands of DNA to form a tetrahedron, with one strand of RNA affixed to each of the six edges. The tetrahedron is further equipped with targeting protein, three folate molecules, which lead the DNA nanoparticles to the abundant folate receptors found on some tumors. The result showed that the gene expression targeted by the RNAi, luciferase, dropped by more than half. This study shows promise in using DNA nanotechnology as an effective tool to deliver treatment using the emerging RNA Interference technology.

Design

DNA nanostructures must be rationally designed so that individual nucleic acid strands will assemble into the desired structures. This process usually begins with specification of a desired target structure or function. Then, the overall secondary structure of the target complex is determined, specifying the arrangement of nucleic acid strands within the structure, and which portions of those strands should be bound to each other.

The last step is the primary structure design, which is the specification of the actual base sequences of each nucleic acid strand.

Structural Design

The first step in designing a nucleic acid nanostructure is to decide how a given structure should be represented by a specific arrangement of nucleic acid strands. This design step determines the secondary structure, or the positions of the base pairs that hold the individual strands together in the desired shape. Several approaches have been demonstrated:

- Tile-based structures. This approach breaks the target structure into smaller units with strong binding between the strands contained in each unit, and weaker interactions between the units. It is often used to make periodic lattices, but can also be used to implement algorithmic self-assembly, making them a platform for DNA computing. This was the dominant design strategy used from the mid-1990s until the mid-2000s, when the DNA origami methodology was developed.

- Folding structures. An alternative to the tile-based approach, folding approaches make the nanostructure from one long strand, which can either have a designed sequence that folds due to its interactions with itself, or it can be folded into the desired shape by using shorter, "staple" strands. This latter method is called DNA origami, which allows forming nanoscale two- and three-dimensional shapes.

- Dynamic assembly. This approach directly controls the kinetics of DNA self-assembly, specifying all of the intermediate steps in the reaction mechanism in addition to the final product. This is done using starting materials which adopt a hairpin structure; these then assemble into the final conformation in a cascade reaction, in a specific order. This approach has the advantage of proceeding isothermally, at a constant temperature. This is in contrast to the thermodynamic approaches, which require a thermal annealing step where a temperature change is required to trigger the assembly and favor proper formation of the desired structure.

Sequence Design

After any of the above approaches are used to design the secondary structure of a target complex, an actual sequence of nucleotides that will form into the desired structure must be devised. Nucleic acid design is the process of assigning a specific nucleic acid base sequence to each of a structure's constituent strands so that they will associate into a desired conformation. Most methods have the goal of designing sequences so that the target structure has the lowest energy, and is thus the most thermodynamically favorable, while incorrectly assembled structures have higher energies and are thus disfavored. This is done either through simple, faster heuristic methods such as sequence symmetry minimization, or by using a full nearest-neighbor thermodynamic model, which is more accurate but slower and more computationally intensive. Geometric models are used to examine tertiary structure of the nanostructures and to ensure that the complexes are not overly strained.

Nucleic acid design has similar goals to protein design. In both, the sequence of monomers is designed to favor the desired target structure and to disfavor other structures. Nucleic acid design has the advantage of being much computationally easier than protein design, because the simple base pairing rules are sufficient to predict a structure's energetic favorability, and detailed information about the overall three-dimensional folding of the structure is not required. This allows the use of simple heuristic methods that yield experimentally robust designs. However, nucleic acid structures are less versatile than proteins in their function because of proteins' increased ability to fold into complex structures, and the limited chemical diversity of the four nucleotides as compared to the twenty proteinogenic amino acids.

Materials and Methods

The sequences of the DNA strands making up a target structure are designed computationally, using molecular modeling and thermodynamic modeling software. The nucleic acids themselves are then synthesized using standard oligonucleotide synthesis methods, usually automated in an oligonucleotide synthesizer, and strands of custom sequences are commercially available. Strands can be purified by denaturing gel electrophoresis if needed, and precise concentrations determined via any of several nucleic acid quantitation methods using ultraviolet absorbance spectroscopy.

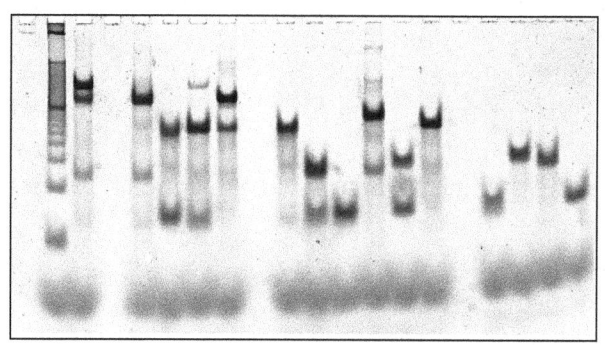

Gel electrophoresis methods, such as this formation assay on a DX complex, are used to ascertain whether the desired structures are forming properly. Each vertical lane contains a series of bands, where each band is characteristic of a particular reaction intermediate.

The fully formed target structures can be verified using native gel electrophoresis, which gives size and shape information for the nucleic acid complexes. An electrophoretic mobility shift assay can assess whether a structure incorporates all desired strands. Fluorescent labeling and Förster resonance energy transfer (FRET) are sometimes used to characterize the structure of the complexes.

Nucleic acid structures can be directly imaged by atomic force microscopy, which is well suited to extended two-dimensional structures, but less useful for discrete three-dimensional structures because of the microscope tip's interaction with the fragile nucleic acid structure; transmission electron microscopy and cryo-electron microscopy are often used in this case. Extended three-dimensional lattices are analyzed by X-ray crystallography.

History

The conceptual foundation for DNA nanotechnology was first laid out by Nadrian Seeman in the early 1980s. Seeman's original motivation was to create a three-dimensional DNA lattice for orienting other large molecules, which would simplify their crystallographic study by eliminating the difficult process of obtaining pure crystals. This idea had reportedly come to him in late 1980, after realizing the similarity between the woodcut *Depth* by M. C. Escher and an array of DNA six-arm junctions. Several natural branched DNA structures were known at the time, including the DNA replication fork and the mobile Holliday junction, but Seeman's insight was that immobile nucleic acid junctions could be created by properly designing the strand sequences to remove symmetry in the assembled molecule, and that these immobile junctions could in principle be combined into rigid crystalline lattices. The first theoretical paper proposing this scheme was published in 1982, and the first experimental demonstration of an immobile DNA junction was published the following year.

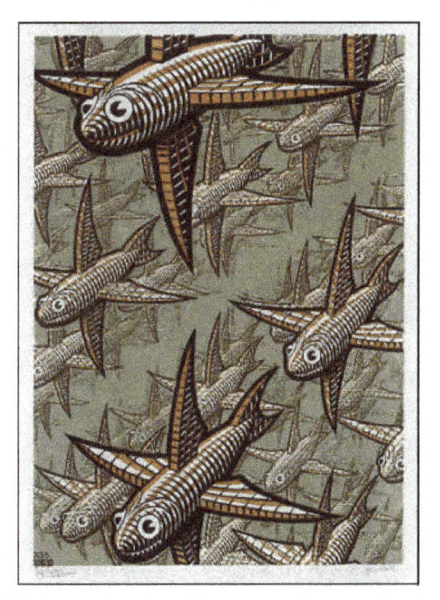

Three-dimensional lattices of DNA to orient hard-to-crystallize molecules. This led to the beginning of the field of DNA nanotechnology.

In 1991, Seeman's laboratory published a report on the synthesis of a cube made of DNA, the first synthetic three-dimensional nucleic acid nanostructure, for which he received the 1995 Feynman Prize in Nanotechnology. This was followed by a DNA truncated octahedron. However, it soon became clear that these structures, polygonal shapes with flexible junctions as their vertices, were not rigid enough to form extended three-dimensional lattices. Seeman developed the more rigid double-crossover (DX) structural motif, and in 1998, in collaboration with Erik Winfree, published the creation of two-dimensional lattices of DX tiles. These tile-based structures had the advantage that they provided the ability to implement DNA computing, which was demonstrated by Winfree and Paul Rothemund in their 2004 paper on the algorithmic self-assembly of a Sierpinski gasket structure, and for which they shared the 2006 Feynman Prize in Nanotechnology. Winfree's key insight was that the DX tiles could be used as Wang tiles, meaning that their assembly could perform computation. The synthesis of a three-dimensional lattice was finally published by Seeman in 2009, nearly thirty years after he had set out to achieve it.

New abilities continued to be discovered for designed DNA structures throughout the 2000s. The first DNA nanomachine—a motif that changes its structure in response to an input—was demonstrated in 1999 by Seeman. An improved system, which was the first nucleic acid device to make use of toehold-mediated strand displacement, was demonstrated by Bernard Yurke the following year. The next advance was to translate this into mechanical motion, and in 2004 and 2005, several DNA walker systems were demonstrated by the groups of Seeman, Niles Pierce, Andrew Turberfield, and Chengde Mao. The idea of using DNA arrays to template the assembly of other molecules such as nanoparticles and proteins, first suggested by Bruche Robinson and Seeman in 1987, was demonstrated in 2002 by Seeman, Kiehl et al. and subsequently by many other groups.

In 2006, Rothemund first demonstrated the DNA origami method for easily and robustly forming folded DNA structures of arbitrary shape. Rothemund had conceived of this method as being conceptually intermediate between Seeman's DX lattices, which used many short strands, and William Shih's DNA octahedron, which consisted mostly of one very long strand. Rothemund's DNA origami contains a long strand which folding is assisted by several short strands. This method allowed forming much larger structures than formerly possible, and which are less technically demanding to design and synthesize. DNA origami was the cover story of *Nature* on March 15, 2006. Rothemund's research demonstrating two-dimensional DNA origami structures was followed by the demonstration of solid three-dimensional DNA origami by Douglas *et al.* in 2009, while the labs of Jørgen Kjems and Yan demonstrated hollow three-dimensional structures made out of two-dimensional faces.

DNA nanotechnology was initially met with some skepticism due to the unusual non-biological use of nucleic acids as materials for building structures and doing computation, and the preponderance of proof of principle experiments that extended the abilities of the field but were far from actual applications. Seeman's 1991 paper on the synthesis of the DNA cube was rejected by the journal *Science* after one reviewer praised its originality while another criticized it for its lack of biological relevance. By the early

2010s, however, the field was considered to have increased its abilities to the point that applications for basic science research were beginning to be realized, and practical applications in medicine and other fields were beginning to be considered feasible. The field had grown from very few active laboratories in 2001 to at least 60 in 2010, which increased the talent pool and thus the number of scientific advances in the field during that decade.

Self-assembled Monolayer

Self-assembled monolayers (SAM) of organic molecules are molecular assemblies formed spontaneously on surfaces by adsorption and are organized into more or less large ordered domains. In some cases molecules that form the monolayer do not interact strongly with the substrate. This is the case for instance of the two-dimensional supramolecular networks of e.g. Perylene-tetracarboxylicacid-dianhydride (PTCDA) on gold or of e.g. porphyrins on highly oriented pyrolitic graphite (HOPG). In other cases the molecules possess a head group that has a strong affinity to the substrate and anchors the molecule to it. Such a SAM consisting of a head group, tail and functional end group is depicted in Figure 1. Common head groups include thiols, silanes, phosphonates, etc.

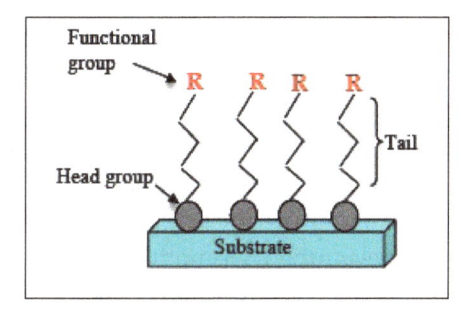

Figure 1. Representation of a SAM structure.

SAMs are created by the chemisorption of "head groups" onto a substrate from either the vapor or liquid phase followed by a slow organization of "tail groups". Initially, at small molecular density on the surface, adsorbate molecules form either a disordered mass of molecules or form an ordered two-dimensional "lying down phase", and at higher molecular coverage, over a period of minutes to hours, begin to form three-dimensional crystalline or semicrystalline structures on the substrate surface. The "head groups" assemble together on the substrate, while the tail groups assemble far from the substrate. Areas of close-packed molecules nucleate and grow until the surface of the substrate is covered in a single monolayer.

Adsorbate molecules adsorb readily because they lower the surface free-energy of the substrate and are stable due to the strong chemisorption of the "head groups." These bonds create monolayers that are more stable than the physisorbed bonds of Langmuir–Blodgett films. A Trichlorosilane based "head group", for example in a FDTS molecule reacts with an hydroxyl group on a substrate, and forms very stable, covalent bond [R-Si-O-substrate] with an energy of 452 kJ/mol. Thiol-metal bonds, that are on the order of 100 kJ/mol, making the bond a fairly stable in a variety of temperature, solvents, and potentials. The monolayer packs tightly due to van der Waals interactions, thereby reducing its own free energy. The adsorption can be described by the Langmuir adsorption isotherm if lateral interactions are neglected. If they cannot be neglected, the adsorption is better described by the Frumkin isotherm.

Types

Selecting the type of head group depends on the application of the SAM. Typically, head groups are connected to a molecular chain in which the terminal end can be functionalized (i.e. adding –OH, –NH2, –COOH, or –SH groups) to vary the wetting and interfacial properties. An appropriate substrate is chosen to react with the head group. Substrates can be planar surfaces, such as silicon and metals, or curved surfaces, such as nanoparticles. Alkanethiols are the most commonly used molecules for SAMs. Alkanethiols are molecules with an alkyl chain, $(C-C)^n$ chain, as the back bone, a tail group, and a S-H head group. Other types of interesting molecules include aromatic thiols, of interest in molecular electronics, in which the alkane chain is (partly) replaced by aromatic rings. An example is the dithiol 1,4-Benzenedimethanethiol ($SHCH_2C_6H_4CH_2SH$)). Interest in such dithiols stems from the possibility of linking the two sulfur ends to metallic contacts, which was first used in molecular conduction measurements. Thiols are frequently used on noble metal substrates because of the strong affinity of sulfur for these metals. The sulfur gold interaction is semi-covalent and has a strength of approximately 45kcal/mol. In addition, gold is an inert and biocompatible material that is easy to acquire. It is also easy to pattern via lithography, a useful feature for applications in nanoelectromechanical systems (NEMS). Additionally, it can withstand harsh chemical cleaning treatments. Recently other chalcogenide SAMs: selenides and tellurides have attracted attention in a search for different bonding characteristics to substrates affecting the SAM characteristics and which could be of interest in some applications such as molecular electronics. Silanes are generally used on nonmetallic oxide surfaces; however monolayers formed from covalent bonds between silicon and carbon or oxygen cannot be considered self assembled because they do not form reversibly. Self-assembled monolayers of thiolates on noble metals are a special case because the metal-metal bonds become reversible after the formation of the thiolate-metal complex. This reversibility is what gives rise to vacancy islands and it is why SAMs of alkanethiolates can be thermally desorbed and undergo exchange with free thiols.

Preparation

Metal substrates for use in SAMs can be produced through physical vapor deposition techniques, electrodeposition or electroless deposition. Thiol or selenium SAMs produced by adsorption from solution are typically made by immersing a substrate into a dilute solution of alkane thiol in ethanol, though many different solvents can be used besides use of pure liquids. While SAMs are often allowed to form over 12 to 72 hours at room temperature, SAMs of alkanethiolates form within minutes. Special attention is essential in some cases, such as that of dithiol SAMs to avoid problems due to oxidation or photoinduced processes, which can affect terminal groups and lead to disorder and multilayer formation. In this case appropriate choice of solvents, their degassing by inert gasses and preparation in the absence of light is crucial and allows formation of "standing up" SAMs with free –SH groups. Self-assembled monolayers can also be adsorbed from the vapor phase. In some cases when obtaining an ordered assembly is difficult or when different density phases need to be obtained substitutional self-assembly is used. Here one first forms the SAM of a given type of molecules, which give rise to ordered assembly and then a second assembly phase is performed (e.g. by immersion into a different solution). This method has also been used to give information on relative binding strengths of SAMs with different head groups and more generally on self-assembly characteristics.

Characterization

The thicknesses of SAMs can be measured using ellipsometry and X-ray photoelectron spectroscopy (XPS), which also give information on interfacial properties. The order in the SAM and orientation of molecules can be probed by Near Edge Xray Absorption Fine Structure (NEXAFS) and Fourier Transform Infrared Spectroscopy in Reflection Absorption Infrared Spectroscopy (RAIRS) studies. Numerous other spectroscopic techniques are used such as Second-harmonic generation (SHG), Sum-frequency generation (SFG), Surface-enhanced Raman scattering (SERS), as well as High-resolution electron energy loss spectroscopy (HREELS). The structures of SAMs are commonly determined using scanning probe microscopy techniques such as atomic force microscopy (AFM) and scanning tunneling microscopy (STM). STM has been able to help understand the mechanisms of SAM formation as well as determine the important structural features that lend SAMs their integrity as surface-stable entities. In particular STM can image the shape, spatial distribution, terminal groups and their packing structure. AFM offers an equally powerful tool without the requirement of the SAM being conducting or semi-conducting. AFM has been used to determine chemical functionality, conductance, magnetic properties, surface charge, and frictional forces of SAMs. More recently, however, diffractive methods have also been used. The structure can be used to characterize the kinetics and defects found on the monolayer surface. These techniques have also shown physical differences between SAMs with planar substrates and nanoparticle substrates. An alternative characterisation instrument for measuring the self-assembly in real time is dual polarisation interferometry where the refractive index, thickness, mass and birefringence of the self assembled layer are quantified at high resolution. Contact angle measurements can be used to determine the surface free-energy which reflects the average composition of the surface of the SAM and can be used to probe the kinetics and thermodynamics of the formation of SAMs. The kinetics of adsorption and temperature induced desorption as well as information on structure can also be obtained in real time by ion scattering techniques such as low energy ion scattering (LEIS) and time of flight direct recoil spectroscopy (TOFDRS).

Defects

Defects due to both external and intrinsic factors may appear. External factors include the cleanliness of the substrate, method of preparation, and purity of the adsorbates. SAMs intrinsically form defects due to the thermodynamics of formation, e.g. thiol SAMs on gold typically exhibit etch pits (monatomic vacancy islands) likely due to extraction of adatoms from the substrate and formation of adatom-adsorbate moieties. Recently, a new type of fluorosurfactants have found that can form nearly perfect monolayer on gold substrate due to the increase of mobility of gold surface atoms.

Nanoparticle Properties

The structure of SAMs is also dependent on the curvature of the substrate. SAMs on nanoparticles, including colloids and nanocrystals, "stabilize the reactive surface of the particle and present organic functional groups at the particle-solvent interface". These organic functional groups are useful for applications, such as immunoassays, that are dependent on chemical composition of the surface.

Kinetics

There is evidence that SAM formation occurs in two steps: an initial fast step of adsorption and a second slower step of monolayer organization. Adsorption occurs at the liquid–liquid, liquid–vapor, and liquid-solid interfaces. The transport of molecules to the surface occurs due to a combination of diffusion and convective transport. According to the Langmuir or Avrami kinetic model the rate of deposition onto the surface is proportional to the free space of the surface.

$$\mathbf{k}(1-\theta) = \frac{d\theta}{dt}.$$

Where θ is the proportional amount of area deposited and k is the rate constant. Although this model is robust it is only used for approximations because it fails to take into account intermediate processes. Dual polarisation interferometry being a real time technique with ~10 Hz resolution can measure the kinetics of monolayer self-assembly directly.

Once the molecules are at the surface the self-organization occurs in three phases:

1. A low-density phase with random dispersion of molecules on the surface.
2. An intermediate-density phase with conformational disordered molecules or molecules lying flat on the surface.

3. A high-density phase with close-packed order and molecules standing normal to the substrate's surface.

The phase transitions in which a SAM forms depends on the temperature of the environment relative to the triple point temperature, the temperature in which the tip of the low-density phase intersects with the intermediate-phase region. At temperatures below the triple point the growth goes from phase 1 to phase 2 where many islands form with the final SAM structure, but are surrounded by random molecules. Similar to nucleation in metals, as these islands grow larger they intersect forming boundaries until they end up in phase 3.

At temperatures above the triple point the growth is more complex and can take two paths. In the first path the heads of the SAM organize to their near final locations with the tail groups loosely formed on top. Then as they transit to phase 3, the tail groups become ordered and straighten out. In the second path the molecules start in a lying down position along the surface. These then form into islands of ordered SAMs, where they grow into phase 3.

The nature in which the tail groups organize themselves into a straight ordered monolayer is dependent on the inter-molecular attraction, or Van der Waals forces, between the tail groups. To minimize the free energy of the organic layer the molecules adopt conformations that allow high degree of Van der Waals forces with some hydrogen bonding. The small size of the SAM molecules are important here because Van der Waals forces arise from the dipoles of molecules and are thus much weaker than the surrounding surface forces at larger scales. The assembly process begins with a small group of molecules, usually two, getting close enough that the Van der Waals forces overcome the surrounding force. The forces between the molecules orient them so they are in their straight, optimal, configuration. Then as other molecules come close by they interact with these already organized molecules in the same fashion and become a part of the conformed group. When this occurs across a large area the molecules support each other into forming their SAM shape seen in Figure 1. The orientation of the molecules can be described with two parameters: α and β. α is the angle of tilt of the backbone from the surface normal. In typical applications α varies from 0 to 60 degrees depending on the substrate and type of SAM molecule. β is the angle of rotation along the long axis of tee molecule. β is usually between 30 and 40 degrees. In some cases existence of kinetic traps hindering the final ordered orientation has been pointed out. Thus in case of dithiols formation of a "lying down" phase was considered an impediment to formation of "standing up" phase, however various recent studies indicate this is not the case.

Many of the SAM properties, such as thickness, are determined in the first few minutes. However, it may take hours for defects to be eliminated via annealing and for final SAM properties to be determined. The exact kinetics of SAM formation depends on the adsorbate, solvent and substrate properties. In general, however, the kinetics are dependent on both preparations conditions and material properties of the solvent, adsorbate and substrate. Specifically, kinetics for adsorption from a liquid solution are dependent on:

- Temperature – room-temperature preparation improves kinetics and reduces defects.

- Concentration of adsorbate in the solution – low concentrations require longer immersion times and often create highly crystalline domains.

- Purity of the adsorbate – impurities can affect the final physical properties of the SAM.

- Dirt or contamination on the substrate – imperfections can cause defects in the SAM.

The final structure of the SAM is also dependent on the chain length and the structure of both the adsorbate and the substrate. Steric hindrance and metal substrate properties, for example, can affect the packing density of the film, while chain length affects SAM thickness. Longer chain length also increases the thermodynamic stability.

Patterning

1. Locally Attract

This first strategy involves locally depositing self-assembled monolayers on the surface only where the nanostructure will later be located. This strategy is advantageous because it involves high throughput methods that generally involve fewer steps than the other two strategies. The major techniques that use this strategy are:

- Micro-contact printing

    Micro-contact printing or soft lithography is analogous to printing ink with a rubber stamp. The SAM molecules are inked onto a pre-shaped elastomeric stamp with a solvent and transferred to the substrate surface by stamping. The SAM solution is applied to the entire stamp but only areas that make contact with the surface allow transfer of the SAMs. The transfer of the SAMs is a complex diffusion process that depends on the type of molecule, concentration, duration of contact, and pressure applied. Typical stamps use PDMS because its elastomeric properties, $E = 1.8$ MPa, allow it to fit the countour of micro surfaces and its low surface energy, $\gamma = 21.6$ dyn/cm^2. This is a parallel process and can thus place nanoscale objects over a large area in a short time.

- Dip-pen nanolithography
    Dip-pen nanolithography is a process that uses an atomic force microscope to

transfer molecules on the tip to a substrate. Initially the tip is dipped into a reservoir with an ink. The ink on the tip evaporates and leaves the desired molecules attached to the tip. When the tip is brought into contact with the surface a water meniscus forms between the tip and the surface resulting in the diffusion of molecules from the tip to the surface. These tips can have radii in the tens of nanometers, and thus SAM molecules can be very precisely deposited onto a specific location of the surface. This process was discovered by Chad Mirkin and co-workers at Northwestern University.

2. Locally Remove

The locally remove strategy begins with covering the entire surface with a SAM. Then individual SAM molecules are removed from locations where the deposition of nanostructures is not desired. The end result is the same as in the locally attract strategy, the difference being in the way this is achieved. The major techniques that use this strategy are:

- Scanning tunneling microscope

  The scanning tunneling microscope can remove SAM molecules in many different ways. The first is to remove them mechanically by dragging the tip across the substrate surface. This is not the most desired technique as these tips are expensive and dragging them causes a lot of wear and reduction of the tip quality. The second way is to degrade or desorb the SAM molecules by shooting them with an electron beam. The scanning tunneling microscope can also remove SAMs by field desorption and field enhanced surface diffusion.

- Atomic force microscope

  The most common use of this technique is to remove the SAM molecules in a process called shaving, where the atomic force microscope tip is dragged along the surface mechanically removing the molecules. An atomic force microscope can also remove SAM molecules by local oxidation nanolithography.

- Ultraviolet irradiation

  In this process, UV light is projected onto the surface with a SAM through a pattern of apperatures in a chromium film. This leads to photo oxidation of the SAM molecules. These can then be washed away in a polar solvent. This process has 100 nm resolutions and requires exposure time of 15–20 minutes.

3. Modify Tail Groups

The final strategy focuses not on the deposition or removal of SAMS, but the modification of terminal groups. In the first case the terminal group can be modified to remove functionality so that SAM molecule will be inert. In the same regards the terminal group can be modified to add functionality so it can accept different materials or have different properties than the original SAM terminal group. The major techniques that use this strategy are:

- Focused electron beam and ultraviolet irradiation

    Exposure to electron beams and UV light changes the terminal group chemistry. Some of the changes that can occur include the cleavage of bonds, the forming of double carbon bonds, cross-linking of adjacent molecules, fragmentation of molecules, and confromational disorder.

- Atomic force microscope

    A conductive AFM tip can create an electrochemical reaction that can change the terminal group.

Applications

Thin-film SAMs

SAMs are an inexpensive and versatile surface coating for applications including control of wetting and adhesion, chemical resistance, bio compatibility, sensitization, and molecular recognition for sensors and nano fabrication. Areas of application for SAMs include biology, electrochemistry and electronics, nanoelectromechanical systems (NEMS) and microelectromechanical systems (MEMS), and everyday household goods. SAMs can serve as models for studying membrane properties of cells and organelles and cell attachment on surfaces. SAMs can also be used to modify the surface properties of electrodes for electrochemistry, general electronics, and various NEMS and MEMS. For example, the properties of SAMs can be used to control electron transfer in electrochemistry. They can serve to protect metals from harsh chemicals and etchants. SAMs can also reduce sticking of NEMS and MEMS components in humid environments. In the same way, SAMs can alter the properties of glass. A common household product, Rain-X, utilizes SAMs to create a hydrophobic monolayer on car windshields to keep them clear of rain. Another application is an anti-adhesion coating on nanoimprint lithography (NIL) tools and stamps. One can also coat injection molding tools for polymer replication with a Perfluordecyltrichlorosilane SAM.

Thin film SAMs can also be placed on nanostructures. In this way they functionalize the nanostructure. This is advantageous because the nanostructure can now selectively attach itself to other molecules or SAMs. This technique is useful in biosensors or other MEMS devices that need to separate one type of molecule from its environment. One example is the use of magnetic nanoparticles to remove a fungus from a blood stream. The nanoparticle is coated with a SAM that binds to the fungus. As the contaminated blood is filtered through a MEMS device the magnetic nanoparticles are inserted into the blood where they bind to the fungus and are then magnetically driven out of the blood stream into a nearby laminar waste stream.

Patterned SAMs

SAMs are also useful in depositing nanostructures, because each adsorbate molecule can be tailored to attract two different materials. Current techniques utilize the head to attract to a surface, like a plate of gold. The terminal group is then modified to attract a specific material like a particular nanoparticle, wire, ribbon, or other nanostructure. In this way, wherever the a SAM is patterned to a surface there will be nanostructures attached to the tail groups. One example is the use of two types of SAMs to align single wall carbon nanotubes, SWNTs. Dip pen nanolithography was used to pattern a 16-mercaptohexadecanoic acid (MHA)SAM and the rest of the surface was passivated with 1-octadecanethiol (ODT) SAM. The polar solvent that is carrying the SWNTs is attracted to the hydrophilic MHA; as the solvent evaporates, the SWNTs are close enough to the MHA SAM to attach to it due to Van der Waals forces. The nanotubes thus line up with the MHA-ODT boundary. Using this technique Chad Mirkin, Schatz and their co-workers were able to make complex two-dimensional shapes, a representation of a shape created is shown to the right. Another application of patterned SAMs is the functionalization of biosensors. The tail groups can be modified so they have an affinity for cells, proteins, or molecules. The SAM can then be placed onto a biosensor so that binding of these molecules can be detected. The ability to pattern these SAMs allows them to be placed in configurations that increase sensitivity and do not damage or interfere with other components of the biosensor.

Metal Organic Superlattices

There has been considerable interest in use of SAMs for new materials e.g. via formation of two- or three-dimensional metal organic superlattices by assembly of SAM capped nanoparticles or layer by layer SAM-nanoparticle arrays using dithiols.

Supramolecular Assembly

A supramolecular assembly or "supermolecule" is a well defined complex of molecules held together by noncovalent bonds. While a supramolecular assembly can be simply composed of two molecules (e.g., a DNA double helix or an inclusion compound), it is more often used to denote larger complexes of molecules that form sphere-, rod-, or sheet-like species. Micelles, liposomes and biological membranes are examples of supramolecular assemblies. The dimensions of supramolecular assemblies can range from nanometers to micrometers. Thus they allow access to nanoscale objects using a bottom-up approach in far fewer steps than a single molecule of similar dimensions.

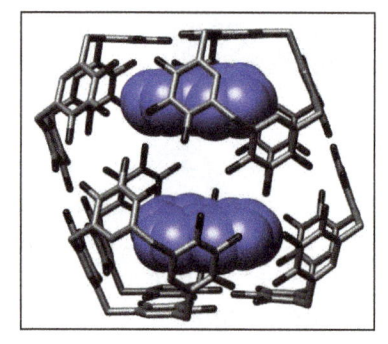

An example of a supramolecular assembly reported by Atwood and coworkers in Science.

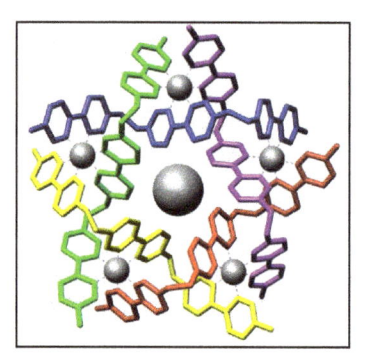

An example of a supramolecular assembly reported by Jean-Marie Lehn and coworkers in Angew.

The process by which a supramolecular assembly forms is called molecular self-assembly. Some try to distinguish self-assembly as the process by which individual molecules form the defined aggregate. Self-organization, then, is the process by which those aggregates create higher-order structures. This can become useful when talking about liquid crystals and block copolymers.

Templating Reactions

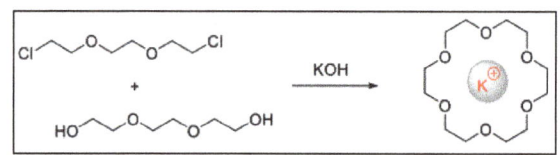

18-crown-6 can be synthesized from using potassium ion as the template cation.

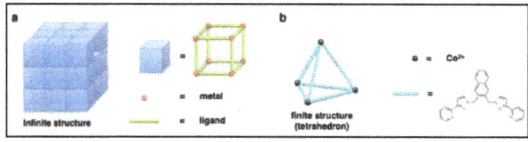

Illustrations of a. metal-organic frameworks and b. supramolecular coordination complexes.

As studied in coordination chemistry, metals (usually transition metals) exist in solution bound to ligands, In many cases, the coordination sphere defines geometries conducive to reactions either between ligands or involving ligands and other external reagents.

A well known metal-templating was described by Charles Pederson in his synthesis of various crown ethers using metal cations as template. For example, 18-crown-6 strongly coordinates potassium ion thus can be prepared through the Williamson ether synthesis using potassium ion as the template metal.

Metals are frequently used for assembly of large supramolecular structures. Metal organic frameworks (MOFs) are one example. MOFs are infinite structures where metal serve as nodes to connect organic ligands together. SCCs are discrete systems where selected metals and ligands undergo self-assembly to form finite supramolecular complexes, usually the size and structure of the complex formed can be determined by the angularity of chosen metal-ligand bonds.

Hydrogen Bond Assisted Supramolecular Assembly

Hydrogen bond-assisted supramolecular assembly is the process of assembling small organic molecules to form large supramolecular structures by non-covalent hydrogen bonding interactions. The directionality, reversibility, and strong bonding nature of hydrogen bond make it an attractive and useful approach in supramolecular assembly. Functional groups such as carboxylic acids, ureas, amines, and amides are commonly used to assemble higher order structures upon hydrogen bonding.

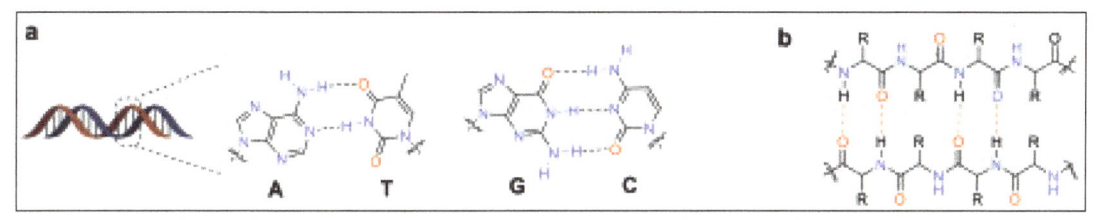

Hydrogen bonds in (a) DNA duplex formation and (b) protein β-sheet structure
(a) Representative hydrogen bond patterns in supramolecular assembly. (b) Hydrogen bond network in cyanuric acid-melamine crystals.

Hydrogen bond play an essential role in the assembly of secondary and tertiary structures of large biomolecules. DNA double helix is formed by hydrogen bonding between nucleobases: adenine and thymine forms two hydrogen bonds, while guanine and cytosine forms three hydrogen bonds (Figure "Hydrogen bonds in (a) DNA duplex formation"). Another prominent example of hydrogen bond-assisted assembly in nature is the formation of protein secondary structures. Both the α-helix and β-sheet are formed through hydrogen bonding between the amide hydrogen and the amide carbonyl oxygen (Figure "Hydrogen bonds in (b) protein β-sheet structure").

In supramolecular chemistry, hydrogen bonds have been broadly applied to crystal engineering, molecular recognition, and catalysis. Hydrogen bonds are among the mostly used synthons in bottom-up approach to engineering molecular interactions in crystals. Representative hydrogen bond patterns for supramolecular assembly is shown in Figure "Representative hydrogen bond patterns in supramolecular assembly". A

1: 1 mixture of cyanuric acid and melamine forms crystal with a highly dense hydrogen-bonding network. This supramolecular aggregates has been used as templates to engineering other crystal structures.

Applications

Supramolecular assemblies have no specific applications but are the subject of many intriguing reactions. A supramolecular assembly of peptide amphiphiles in the form of nanofibers has been shown to promote the growth of neurons. An advantage to this supramolecular approach is that the nanofibers will degrade back into the individual peptide molecules that can be broken down by the body. By self-assembling of dendritic dipeptides, hollow cylinders can be produced. The cylindrical assemblies possess internal helical order and self-organize into columnar liquid crystalline lattices. When inserted into vesicular membranes, the porous cylindrical assemblies mediate transport of protons across the membrane. Self-assembly of dendrons generates arrays of nanowires. Electron donor-acceptor complexes comprise the core of the cylindrical supramolecular assemblies, which further self-organize into two-dimensional columnar liquid crystaline lattices. Each cylindrical supramolecular assembly functions as an individual wire. High charge carrier mobilities for holes and electrons were obtained.

References

- Background: Pelesko, John A. (2007). Self-assembly: the science of things that put themselves together. New York: Chapman & Hall/CRC. pp. 5, 7. ISBN 978-1-58488-687-7
- Wang, Ji-Tao (2002). Nonequilibrium Nondissipative Thermodynamics: With Application to Low-pressure Diamond Synthesis. Berlin: Springer Verlag. p. 65. ISBN 978-3-540-42802-2
- Ariga, Katsuhiko; Kunitake, Toyoki (2006). Supramolecular chemistry - fundamentals and appl - cations.(English ed.). Berlin [u.a.]: Springer. ISBN 3540012982. Retrieved 04, April 2020
- Lu, Yicheng; Zhong, Jian (2004). Todd Steiner, ed. Semiconductor Nanostructures for Optoelectronic Applications. Norwood, MA: Artech House, Inc. pp. 191–192. ISBN 978-1-58053-751-3
- Applications: Rietman, Edward A. (2001). Molecular engineering of nanosystems. Springer. pp. 209–212. ISBN 978-0-387-98988-4. Retrieved 17, April 2020
- Methods: Gallagher, S. R.; Desjardins, P. (1 July 2011). "Quantitation of nucleic acids and proteins". Current Protocols Essential Laboratory Techniques. doi:10.1002/9780470089941. et0202s5. ISBN 0470089938
- History: Pelesko, John A. (2007). Self-assembly: the science of things that put themselves together. New York: Chapman & Hall/CRC. pp. 201, 242, 259. ISBN 978-1-58488-687-7
- Rosen, Milton J. (2004). Surfactants and interfacial phenomena. Hoboken, NJ: Wiley-Interscience ISBN 978-0-471-47818-8

Carbon Nanotubes

The cylindrical molecules made of sheets of single-layer carbon atoms are known as carbon nanotubes. Carbon nanotubes have remarkable electrical conductivity. In order to completely understand carbon nanotube, it is necessary to understand the processes related to it. The following chapter elucidates the varied processes and mechanisms associated with this area of study.

Carbon nanotubes (CNTs) are nanostructures derived from rolled graphene planes and possess various interesting chemical and physical properties, have been extensively used in biomedicine. The discovery of carbon nanotubes by Iijima in 1991 using High Resolution Electron Microscopy (HREM) has stimulated intense experimental and theoretical studies on carbon nanotubes. Carbon nanotubes are allotropes of carbon that have a nanostructure, which can have a length-to diameter ratio more than 1,000,000. Theoretical studies have predicted exciting electronic properties for the nanotubes. The potential application of carbon nanotubes to the synthesis of nanowires has been demonstrated. HREM is a robust approach for the characterization of microstructure and it is most suited to the study of nanotubes, it should be pointed out that the image obtained is a two-dimensional projection of a three-dimensional object. CNTs can be conjugated with various biological molecules including drugs, proteins and nucleic acid to afford bio functionalities. Moreover, the aromatic network existing on the CNT surface allows efficient loading of aromatic molecules such as chemotherapeutic drugs via stacking. The versatile chemistry of carbon nanotubes enables a wide range of their applications in biomedicine. CNTs exist as single (SWNTs), and multi-walled (MWNTs) structures. They present several interesting properties, such as high aspect-ratio, ultra-light weight, tremendous strength, high thermal conductivity and remarkable electronic properties ranging from metallic to semiconducting.

Types of Carbon Nanotubes

Single Wall Carbon Nanotube

Geometrically, there is no restriction on the tube diameter. However, calculations have shown that collapsing the single wall tube into a flattened two-layer ribbon is energetically more favorable than maintaining the tubular morphology beyond a diameter value of ≈2.5nm. On the other hand, it is easy to grasp intuitively that the shorter the radius of curvature, the higher the stress and the energetic cost, although SWNTs with diameters as low as 0.4 nm have been synthesized successfully. A suitable energetic compromise is therefore reached for ≈1.4 nm, the most frequent diameter encountered regardless of the synthesis technique (at least for those based on solid carbon sources) when conditions ensuring high SWNT yields are used. There is no such restriction on the nanotube length, which only depends on the limitations of the preparation method and the specific conditions used for the synthesis (thermal gradients, residence time and so on). Experimental data are consistent with these statements, since SWNTs wider than 2.5 nm are only rarely reported in the literature, whatever the preparation method, while the length of the SWNTs can be in the micrometer or the millimeter range. These features make single-wall carbon nanotubes a unique example of single molecules with huge aspect ratios.

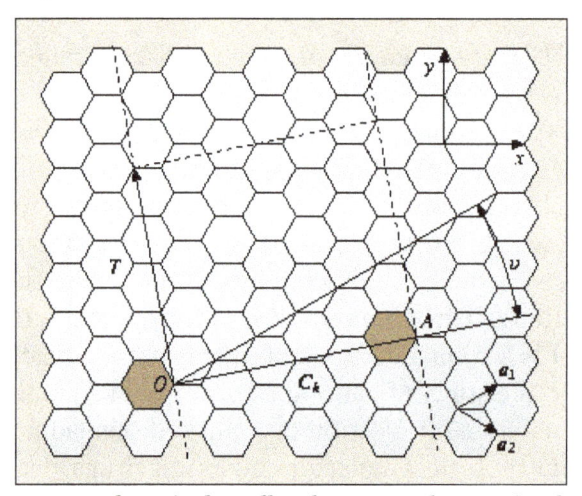

Figure: Sketch of the way to make a single-wall carbon nanotube, starting from a graphene sheet.

Two important consequences derive from the SWNT structure as described below:

- All carbon atoms are involved in hexagonal aromatic rings only and are therefore in equivalent positions, except at each nanotube tip, where 6×5 = 30 atoms are involved in pentagonal rings (considering that adjacent pentagons are unlikely) – though not more, not less, as a consequence of Euler's rule that also governs the fullerene structure. For ideal SWNTs, chemical reactivity will therefore be highly favored at the tube tips, at the locations of the pentagonal rings.

- Although carbon atoms are involved in aromatic rings, the C=C bond angles are not planar. This means that the hybridization of carbon atoms is not pure sp^2; it has some degree of the sp^3 character, in a proportion that increases as the tube radius of curvature decreases. The effect is the same as for the C60 fullerene molecules, whose radius of curvature is 0.35 nm, and whose bonds therefore have 10% sp^3. On the one hand, this is believed to make the SWNT surface a bit more reactive than regular, planar graphene, even though it still consists of aromatic ring faces. On the other hand, this somehow induces variable overlapping of energy bands, resulting in unique and versatile electronic behavior.

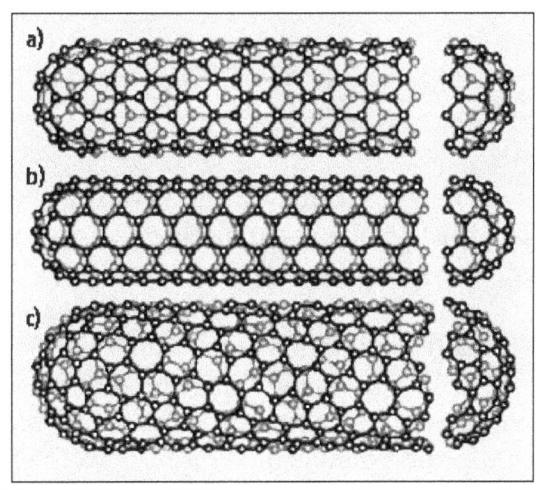

Figure: Sketches of three different SWNT structures that are examples of (a) a zig-zag-type nanotube, (b) an armchair-type nanotube, (c) a helical nanotube.

As illustrated by Figure above, there are many ways to roll a graphene into a single-wall nanotube, with some of the resulting nanotubes possessing planes of symmetry both parallel and perpendicular to the nanotube axis, while others do not. Similar to the terms used for molecules, the latter are commonly called "chiral" nanotubes, since they are unable to be superimposed on their own image in a mirror. "Helical" is however sometimes preferred. The various ways to roll graphene into tubes are therefore mathematically defined by the vector of helicity C_h, and the angle of helicity θ, as follows:

$$OA = C_h = na_1 + ma_2$$

$$a_1 = \frac{a\sqrt{3}}{2}x + \frac{a}{2}y \text{ and } a_2 = \frac{a\sqrt{3}}{2}x - \frac{a}{2}y$$

where $a = 2.46 \text{Å}$

and

$$\cos\theta = \frac{2n+m}{2\sqrt{n^2 + m^2 + nm}}$$

Where n and m are the integers of the vector OA considering the unit vectors a_1 and a_2.

The vector of helicity C_h (= OA) is perpendicular to the tube axis, while the angle of helicity θ is taken with respect to the so-called zig-zag axis: the vector of helicity that results in nanotubes of the "zig-zag" type. The diameter D of the corresponding nanotube is related to C_h by the relation:

$$D = \frac{|C_h|}{\pi} = \frac{a_{cc}\sqrt{3(n^2 + m^2 + mn)}}{\pi}.$$

where,

$$1.41\text{Å} \underset{\text{(graphite)}}{\leq} a_{c=c} \underset{(C_{60})}{\leq} 1.44\text{Å}.$$

The C–C bond length is actually elongated by the curvature imposed by the structure; the average bond length in the C_{60} fullerene molecule is a reasonable upper limit, while the bond length in flat graphene in genuine graphite is the lower limit (corresponding to an infinite radius of curvature). Since C_h, θ, and D are all expressed as a function of the integers n and m, they are sufficient to define any particular SWNT by denoting them (n,m). The values of n and m for a given SWNT can be simply obtained by counting the number of hexagons that separate the extremities of the C_h vector following the unit vector a_1 first and then a_2. In the example of Figure below, the SWNT that is obtained by rolling the graphene so that the two shaded aromatic cycles can be superimposed exactly is a chiral nanotube. Similarly, SWNTs nanotubes respectively, thereby providing examples of zig-zag-type SWNT (with an angle of helicity = 0°), armchair-type SWNT (with an angle of helicity of 30°) and a chiral SWNT, respectively. This also illustrates why the term "chiral" is sometimes inappropriate and should preferably be replaced with "helical". Armchair (n, n) nanotubes, although definitely achiral from the standpoint of symmetry, exhibit a nonzero "chiral angle". "Zig-zag" and "armchair" qualifications for a chiral nanotube refer to the way that the carbon atoms are displayed at the edge of the nanotube cross-section.

Figure: Image of two neighboring chiral SWNTs within a SWNT bundle as seen using high-resolution scanning tunneling microscopy.

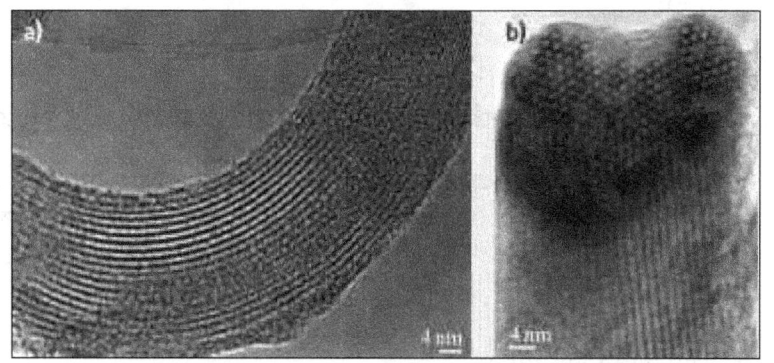

Figure: High-resolution transmission electron microscopy images of a SWNT rope. (a) Longitudinal view. An isolated single SWNT also appears at the top of the image.

Generally speaking, it is clear from above figures that having the vector of helicity perpendicular to any of the three overall C=C bond directions will provide zig-zag-type SWNTs, denoted (n, 0), while having the vector of helicity parallel to one of the three C=C bond directions will provide armchair type SWNTs, denoted (n, n). On the other hand, because of the six fold symmetry of the graphene sheet, the angle of helicity θ for the chiral (n,m) nanotubes is such that $0 < \theta < 30°$. Figure above provides two examples of what chiral SWNTs look like, as seen via atomic force microscopy.

The graphenes in graphite have π electrons which are accommodated by the stacking of graphenes, allowing van der Waals forces to develop. Similar reasons make fullerenes gather and order into fullerite crystals and SWNTs into SWNT ropes. Provided the SWNT diameter distribution is narrow, the SWNTs in ropes tend to spontaneously arrange into hexagonal arrays, which correspond to the highest compactness achievable. This feature brings new periodicities with respect to graphite or turbostratic polyaromatic carbon crystals. Turbostratic structure corresponds to graphenes that are stacked with random rotations or translations instead of being piled up following sequential ABAB positions, as in graphite structure. This implies that no lattice atom plane exists other than the graphene planes themselves (corresponding to the (001) atom plane family). These new periodicities give specific diffraction patterns that are quite different to those of other sp2 -carbon-based crystals, although hk reflections, which account for the hexagonal symmetry of the graphene plane, are still present. On the other hand, 00l reflections, which account for the stacking sequence of graphenes in regular, "multilayered" polyaromatic crystals (which do not exist in SWNT ropes), are absent. This hexagonal packing of SWNTs within the ropes requires that SWNTs exhibit similar diameters, which is the usual case for SWNTs prepared by electric arc or laser vaporization processes. SWNTs prepared using these methods are actually about 1.35 nm wide (diameter of a (10, 10) tube, among others), for reasons that are still unclear but are related to the growth mechanisms specific to the conditions provided by these techniques.

Multiwall Carbon Nanotubes

Building multiwall carbon nanotubes is a little bit more complex, since it involves the various ways graphenes can be displayed and mutually arranged within filamentary morphology. A similar versatility can be expected to the usual textural versatility of polyaromatic solids. Likewise, their diffraction patterns are difficult to differentiate from those of anisotropic polyaromatic solids. The easiest MWNT to imagine is the concentric type (c-MWNT), in which SWNTs with regularly increasing diameters are coaxially arranged (according to a Russian-doll model) into a multiwall nanotube. Such nanotubes are generally formed either by the electric arc technique (without the need for a catalyst), by catalyst-enhanced thermal cracking of mgaseous hydrocarbons, or by CO disproportionation. There can be any number of walls (or coaxial tubes), from two upwards. The inter-tube distance is approximately the same as the inter-graphene distance in turbostratic, polyaromatic solids, 0.34 nm (as opposed to 0.335 nm in genuine graphite), since the increasing radius of curvature imposed on the concentric graphenes prevents the carbon atoms from being arranged as in graphite, with each of the carbon atoms from a graphene facing either a ring center or a carbon atom from the neighboring graphene. However, two cases allow a nanotube to reach – totally or partially – the 3-D crystal periodicity of graphite. One is to consider a high number of concentric graphenes: concentric graphenes with a long radius of curvature. In this case, the shift in the relative positions of carbon atoms from superimposed graphenes is so small with respect to that in graphite that some commensurability is possible. This may result in MWNTs where both structures are associated; in other words they have turbostratic cores and graphitic outer parts. The other case occurs for c-MWNTs exhibiting faceted morphologies, originating either from the synthesis process or more likely from subsequent heat treatment at high temperature (such as 2500 °C) in inert atmosphere. Facets allow the graphenes to resume a flat arrangement of atoms (except at the junction between neighboring facets) which allows the specific stacking sequence of graphite to develop.

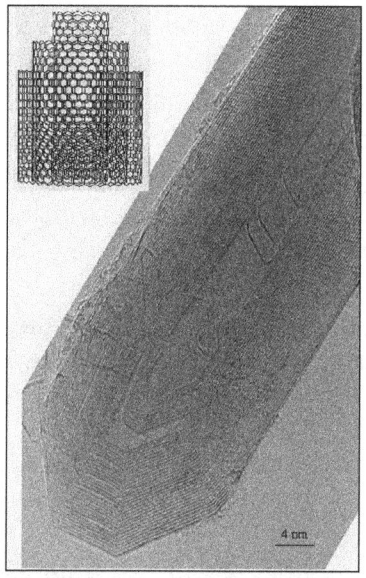

Figure: High-resolution transmission electron microscopy image (longitudinal view) of a concentric multiwall carbon nanotube (c-MWNT) prepared using an electric arc. The insert shows a sketch of the Russian doll-like arrangement of graphenes.

Another frequent inner texture for multiwall carbon nanotubes is the so-called herringbone texture (h-MWNTs), in which the graphenes make an angle with respect to the nanotube axis. The angle value varies upon the processing conditions (such as the catalyst morphology or the composition of the atmosphere), from 0 (in which case the texture becomes that of a c-MWNT) to 90° (in which case the filament is no longer a tube, see below), and the inner diameter varies so that the tubular arrangement can be lost, meaning that the latter are more accurately called nano-fibers rather than nano-tubes. h-MWNTs are exclusively obtained by processes involving catalysts, generally catalyst-enhanced thermal cracking of hydrocarbons or CO disproportionation. One unresolved question is whether the herringbone texture, which actually describes the texture projection rather than the overall three-dimensional texture, originates from the scroll like spiral arrangement of a single graphene ribbon or from the stacking of independent truncated cone-like graphenes in what is also called a "cupstack" texture.

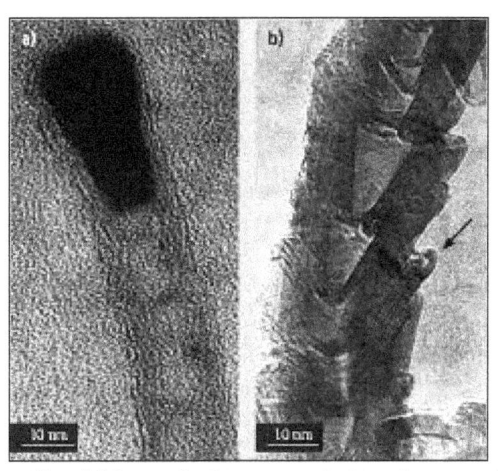

Figure: Some of the earliest high-resolution transmission electron microscopy images of a herringbone (and bamboo) multiwall nanotube (bh-MWNT, longitudinal view) prepared by CO disproportionation on Fe-Co catalyst.

Figure shows (a) As-grown. The nanotube surface is made of free graphene edges. (b) After 2900 °C heat treatment. Both the herringbone and the bamboo textures have become obvious. Graphene edges from the surface have buckled with their neighbors (arrow), closing off access to the inter graphene space.

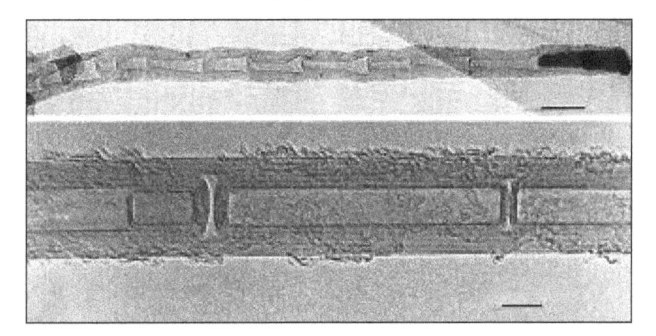

Figure shows transmission electron microscopy images from bamboo multiwall nanotubes (longitudinal views). (a) Low magnification of a bamboo-herringbone multiwall nanotube (bh-MWNT) showing the nearly periodic nature of the texture, which occurs very frequently. (b) High-resolution image of a bamboo-concentric multiwall nanotube (bc-MWNT).

Another common feature is the occurrence, to some degree, of a limited amount of graphenes oriented perpendicular to the nanotube axis, thus forming a "bamboo" texture. This is not a texture that can exist on its own; it affect either the c MWNT (bc-MWNT) or the h-MWNT (bh-MWNT) textures. The question is whether such filaments, although hollow, should still be called nanotubes, since the inner cavity is no longer open all the way along the filament as it is for a genuine tube. These are therefore sometimes referred as "Nano fibers".

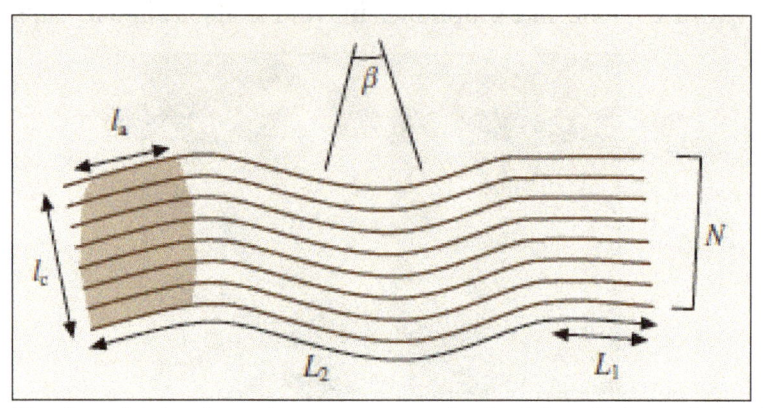

Figure: Sketch explaining the various parameters obtained from high-resolution (lattice fringe mode) transmission electron microscopy, used to quantify Nano-texture.

L_1 is the average length of perfect (distortion free) graphenes of coherent areas; N is the number of piled-up graphenes in coherent (distortion-free) areas; L_2 is the average length of continuous though distorted graphenes within graphene stacks; β is the average distortion angle. L_1 and N are related to the L_a and L_c values obtained from X-ray diffraction.

One Nano-filament that definitely cannot be called a nanotube is built from graphenes oriented perpendicular to the filament axis and stacked as piled-up plates. Although these Nano-filaments actually correspond to hMWNTs with a graphene/MWNT axis angle of 90°, an inner cavity is no longer possible, and such filaments are therefore often referred to as "platelet nano-fibers". Unlike SWNTs, whose aspect ratios are so high that it is almost impossible to find the tube tips, the aspect ratios for MWNTs (and carbon Nano fibers) are generally lower and often allow one to image tube ends by transmission electron microscopy. Aside from c-MWNTs derived from electric arc, which grow in a catalyst-free process, nanotube tips are frequently found to be associated with the catalyst crystals from which they were formed.

Comparison between Single and Multi Wall

Sr. No.	SWNTs	MWNTs
1	Single layer of graphene.	Multiple layer of graphene.
2	Catalyst is required for synthesis.	Can be produced without catalyst.
3	Bulk synthesis is difficult as it requires proper control over growth and atmospheric condition.	Bulk synthesis is easy.
4	Purity is poor.	Purity is high.

5	A chance of defect is more during functionalization.	A chance of defect is less but once occurred it's difficult to improve 6 Less accumulation in body More accumulation in b.
6	Less accumulation in body.	More accumulation in body.
7	Characterization and evaluation is easy .	It has very complex structure.
8	It can be easily twisted and are more pliable.	It cannot be easily twisted.

Synthesis of CNT

High temperature preparation techniques such as arc discharge or laser ablation were first used to produce CNTs however nowadays these methods have been replaced by low temperature chemical vapor deposition (CVD) techniques (<800 °C), since the orientation, alignment, nanotube length, diameter, purity and density of CNTs can be precisely controlled in which technique. Most of these methods require supporting gases and vacuum. However, gas-phase methods are volumetric and hence they are suitable for applications such as composite materials that require large quantities of nanotubes and industrial-scale synthesis methods to make them economically feasible. On the other hand, the disadvantages of gas-phase synthesis methods are low catalyst yields, where only a small percentage of catalysts form nanotubes, short catalyst lifetimes, and low catalyst number density.

During the CNT preparation there are always produced a number of impurities whose type and amount depend on the technique being used. The above mentioned techniques produce powders which contain only a small fraction of CNTs and also other carbonaceous particles such as nanocrystalline graphite, amorphous carbon, fullerenes and different metals (typically Fe, Co, Mo or Ni) that were introduced as catalysts during the synthesis. All these impurities interfere with most of the desired properties of CNTs and cause a serious impediment in characterization and applications. Therefore, one of the most fundamental challenges in CNT science is the development of efficient and simple purification methods. Most common purification methods are based on acid treatment of synthesized CNTs.

Arc Discharge Method

In Arc discharge methods use of higher temperatures (above 1700 °C) for CNT synthesis, which usually causes the growth of CNTs with fewer structural defects in comparison with other techniques. The electric arc method, initially used for producing C60 fullerenes, is the most common and perhaps the easiest way to produce CNTs. MWCNTs were discovered in 1991 by Iijima by the arc-discharge evaporation technique. SWCNTs were produced subsequently in 1993 by the same. In this method, electric arc created between two graphite electrodes leads to an extremely high temperature which is sufficient to sublimate carbon. Either MWCNTs or SWCNTs can he formed when the carbon vapors cools and condenses Generally, MWCNT are formed when there is no catalyst particles between two graphite electrodes; and the SWCNT can be generated by adding Fe, Ni, or Co as catalysts. The catalysts can be introduced by packing metal powder into a hole in the anode. The metal was consumed along with the graphite and created catalyst particles favoring small-diameter SWCNTs.

In case of MWCNTs, the purity and yield depended sensitively on the gas pressure in the reaction vessel. Different atmospheres markedly influence the final morphology of CNTs. They used DC arc discharge of graphite electrodes in helium and methane. By evaporation under high pressured methane gas and high arc current, thick nanotubes embellished with many carbon nanoparticles were obtained. However, thin and long MWNTs were obtained under a methane gas pressure of 50 Torr and an arc current of 20 A for the anode with a diameter of 6 mm. Moreover, it was found that the variation of carbon nanotube morphology was more marked in the case of evaporation in methane gas than that in helium gas.

The SWNTs can be produced when the transition metal catalyst is used. The process of SWNTs growth in arc discharge utilizes a composite anode, usually in hydrogen or argon atmosphere. The anode is made as a composition of graphite and a metal, such as Ni, Fe, Co, Pd, Ag, Pt. etc. or mixtures of Co, Fe, Ni with other elements like Co-Ni, Fe-Ni, Fe-No, Co-Cu, Ni-Cu, Ni-Ti etc. The metal catalyst plays a significant role in the process yield. To ensure high efficiency, the process also needs to be held at a constant gap distance between the electrodes which ensures stable current density and anode consumption rate. In this process, unwanted products such as MWNTs or fullerenes are usually produced too.

Laser Ablation

The laser ablation method uses a pulsed and continuous laser to vaporize a graphite target in an oven, which is filled with helium or argon gas to keep pressure. The laser ablation is similar to the arc discharge, both taking advantage of the very high temperature generated, with the similar optimum background gas and catalyst mix observed. The very similar reaction conditions needed to indicate that the reactions probably occur with the same mechanism for both the laser ablation and electric arc methods. SWNTs were prepared by continuous wave carbon dioxide laser ablation without applying additional heat to the target. They found that the average diameter of SWNTs produced by carbon dioxide laser increased with increasing laser power.

Stramel et al., have successfully applied commercial MWNTs and MWNTs-polystyrene targets (PSNTs) for deposition of composite thin films onto silicon substrates using PLD with a pulsed, diode pumped, Tm: Ho: LuLF laser (a laser host material LuLF (Lu-LiF4) is doped with holmium and thulium in order to reach a laser light production in the vicinity of 2 mm. They found that usage of pure MWNTs targets gives rise to a thin film containing much higher quality MWNTs compared to PSNTs targets. Similarly, prepared MWNTs thin films were deposited by PLD techniques (with Nd: YAG laser) ablating commercially polystyrene-nanotubes pellets on alumina substrates.

Chemical Vapor Deposition (CVD)

Catalytic chemical vapor deposition (CCVD)—either thermal or plasma enhanced (PE)—is now the standard method for the CNTs production. Moreover, there are trends to use other CVD techniques, like water assisted CVD, oxygen assisted CVD, hot-filament (HFCVD), microwave plasma (MPECVD) or radiofrequency CVD (RF-CVD). CCVD is considered to be an economically viable process for large scale and quite pure CNTs production compared with laser ablation. The main advantages of CVD are easy control of the reaction course and high purity of the obtained material, etc. The CNT growth model is still under discussion. Recently, Fotopoulos and Xanthakis discussed the traditionally accepted models, which are base growth and tip growth.

In addition, they mentioned a hypothesis that SWNTs are produced by base growth only, i.e. the cap is formed first and then by a lift off process the CNT is created by addition of carbon atoms at the base. They refer to recent in situ video rate TEM studies which have revealed that the base growth of SWNT in thermal CVD is accompanied by a considerable deformation of the Ni catalyst nanoparticle and the creation of a sub-surface carbon layer. These effects may be produced by the adsorption on the catalyst nanoparticle during pyrolysis. In order to produce SWNTs, the size of the nanoparticle catalyst must be smaller than about 3 nm. The function of the catalyst in the CVD process is the decomposition of carbon source via either plasma irradiation (plasma enhanced CVD, PECVD) or heat (thermal CVD) and its new nucleation to form CNTs. The most frequently used catalysts are transition metals, primarily Fe, Co, or Ni. Sometimes, the traditionally used catalysts are further doped with other metals, e.g. with Au. Concerning the carbon source, the most preferred in CVD are hydrocarbons such as methane, ethane, ethylene, acetylene, xylene, eventually their mixture, isobutane or ethanol. In the case of gaseous carbon source, the CNTs growth efficiency strongly depends on the reactivity and concentration of gas phase intermediates produced together with reactive species and free radicals as a result of hydrocarbon decomposition. Thus, it can be expected that the most efficient intermediates, which have the potential of chemisorption or physisorption on the catalyst surface to initiate CNT growth, should be produced in the gas phase. Commonly used substrates are Ni, Si, SiO_2, Cu, Cu/Ti/Si, stainless steel or glass; rarely $CaCO_3$; graphite and tungsten foil or other substrates were also tested. A special type of substrate, mesoporous silica, was also tested since it might play a templating role in guiding the initial nanotube growth.

Plasma Enhanced CVD

Plasma-enhanced chemical vapor deposition (PECVD) systems have been used to produce both SWNTs and MWNTs. PECVD is a general term, encompassing several differing synthesis methods. In general PECVD can be direct or remote. Direct PECVD systems can be used for the production of MWNT field emitter towers and some SWNTs. A remote PECVD can also be used to produce both MWNTs and SWNTs. For SWNT synthesis in the direct PECVD system, the researchers heated the substrate up to 550-850°C utilized a CH_4-H_2 gas mixture at 500 mT, and applied 900 W of plasma power as well as externally applied magnetic field. The plasma enhanced CVD method generates a glow discharge in a chamber or a reaction furnace by a high frequency voltage applied to both electrodes. A substrate is placed on the grounded electrode. In order to form a uniform film, the reaction gas is supplied from the opposite plate. Catalytic metal, such as Fe, Ni and Co are used on a Si, SiO_2 or glass substrate using thermal CVD or sputtering. As such, PECVD and HWCVD as essentially a crossover between plasma-based growth and CVD synthesis. In contrast, to arc discharge, laser ablation, and solar furnace, the carbon for PECVD synthesis comes from feedstock gases such as CH_4 and CO, so there is no need for a solid graphite source. The argon-assisted plasma is used to break down the feedstock gases into C_2, CH, and other reactive carbon species (CxHy) to facilitate growth at low temperature and pressure.

Functionalization of Carbon Nanotubes

Carbon nanotubes do not disperse in organic matrices due their inert nature and forms bundles with each other. The poor solubility of carbon nanotubes in organic solvents restricts them to be used as drug delivery agents into living systems in drug therapy. Hence many modification approaches like physical, chemical or combined have been exploited for their homogeneous dispersion in common solvents to improve their solubility. The surface of nanotubes can be modified by various ways to enhance their dispersion in organic media. Many applications require covalent modification to meet specific requirements i.e. in case of biosensors the biomolecules require electron mediators to promote electron transfer.

Similarly electrochemical metal ion sensors require specific functional groups which show potential affinity towards particular metal ion. The modification protocol was generally achieved by attaching specific molecule or entity which imparts chemical specificity to the substrate material. These chemical modifications can be easily achieved in many ways.

Covalent Interaction

Covalent modification involves attachment of a functional group onto the carbon nanotube. The functional groups can be attached onto the side wall or ends of the carbon nanotube. The end caps of the carbon nanotubes have the highest reactivity due to its higher pyrimidization angle and the walls of the carbon nanotubes have lower pyrimidization angles which has lower reactivity. Although covalent modifications are very stable, the bonding process disrupts the sp² hybridization of the carbon atoms because an σ-bond is formed. The disruption of the extended sp² hybridization typically decreases the conductance of the carbon nanotubes.

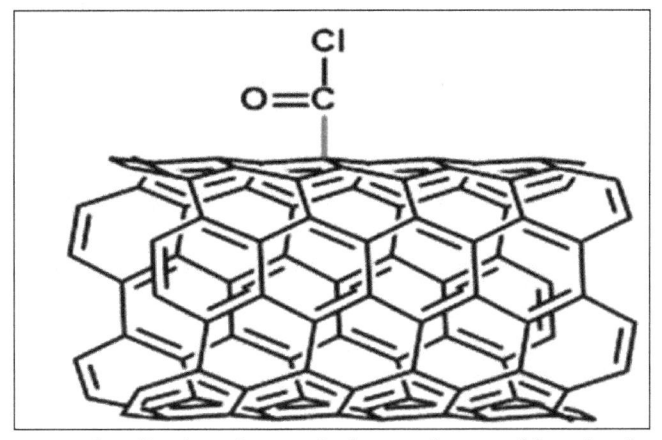

Figure: Functionalization of nanotube by attachment of functional group.

Oxidation

Oxidation of carbon nanotubes also functionalizes the surface by breaking the carbon-carbon bonded network of the Nano layers under acidic conditions. It allows the introduction of oxygen units in the form of carboxyl, phenolic and lactone groups. In liquid-phase reactions, carbon nanotubes are treated with oxidizing solutions of nitric acid or a combination of nitric and sulphuric acid to the same effect. However, over oxidation may occur causing the carbon nanotube to break up into fragments, which are known as carbonaceous fragments.

Esterification/Amidification

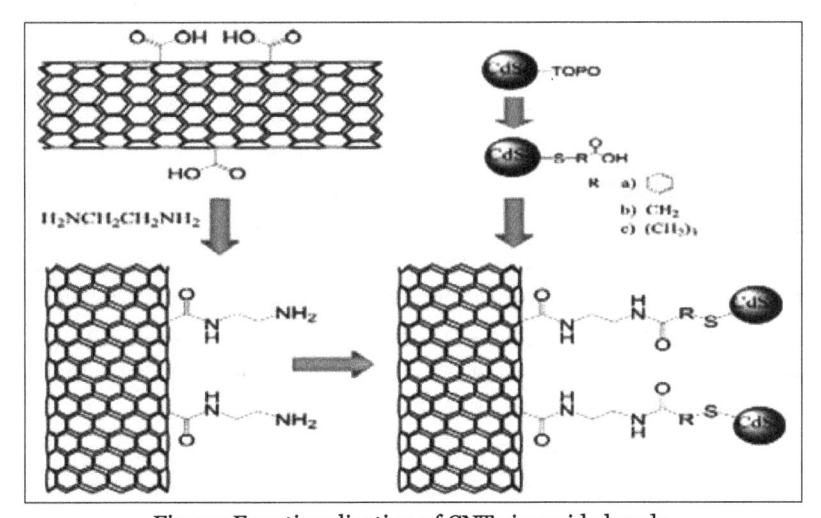

Figure: Functionalization of CNT via amide bonds.

The functionalization of nanotubes is carried out by using carboxylic groups, which acts the precursor for most esterification and admiration reactions. The carboxylic group is converted into an acyl chloride with the use of thionyl or oxalyl chloride which is then reacted with the desired amide, amine, or alcohol. Carbon nanotubes modified with acyl chloride react readily with highly branched molecules such as poly(amindoamine), which acts as a template for silver ion and later being reduced by formaldehyde. Amino-modified carbon nanotubes can be prepared by reacting ethylenediamine with an acyl chloride functionalized carbon nanotubes. In a similar way, thiol stabilized ZnS capped CdSe QDs were protected with 2-aminoethanethiol and linked to the acid terminated carbon nanotubes in presence of a coupling agent like EDC.

Properties of CNT

Mechanical Properties

σ bonding is the strongest in nature, and thus a nanotube that is structured with all σ bonding is regarded as the ultimate fiber with the strength in its tube axis. Both experimental measurements and theoretical calculations agree that a nanotube is as stiff as or stiffer than diamond with the highest Young's modulus and tensile strength. Most theoretical calculations are carried out for perfect structures and give very consistent results. Table below summarizes calculated Young's modulus (tube axis elastic constant) and tensile strength for (10, 10) SWNT and bundle and MWNT with comparison with other materials. The calculation is in agreement with experiments on average. Experimental results show broad discrepancy, especially for MWNTs, because MWNTs contain different amount of defects from different growth approaches. In general, various types of defect-free nanotubes are stronger than graphite. This is mainly because the axial component of σ bonding is greatly increased when a graphite sheet is rolled over to form a seamless cylindrical structure or a SWNT. Young's modulus is independent of tube chirality, but dependent on tube diameter. The highest value is from tube diameter between 1 and 2 nm, about 1 TPa. Large tube is approaching graphite and smaller one is less mechanically stable. When different diameters of SWNTs consist in a coaxial MWNT, the Young's modulus will take the highest value of a SWNT plus contributions from coaxial inter-tube coupling or Van der Waals force. Thus, the Young's modulus for MWNT is higher than a SWNT, typically 1.1 to 1.3 TPa, as determined both experimentally and theoretically. On the other hand, when many SWNTs are held together in a bundle or a rope, the weak Van der Waal force induces a strong shearing among the packed SWNTs. This does not increase but decreases the Young's modulus. It is shown experimentally that the Young's modulus decreases from 1 TPa to 100 GPa when the diameter of a SWNT bundle increases from 3 nm (about 7 (10,10) SWNTs) to 20 nm.

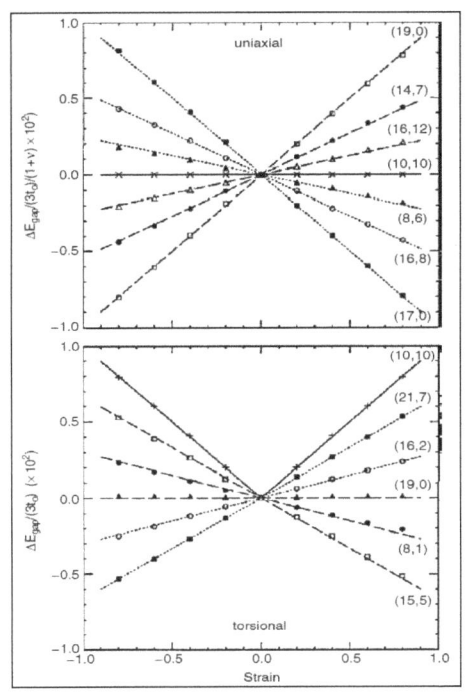

Figure: Band gap change of SWNTs under uniaxial strain (>0 for tension and <0 for compression) and torsional strain (>0) for net bond stretching and <0 for net bond compression).

Table: Mechanical Property of Nanotubes.

	Young's modulus (GPa)	Tensile Strength (GPa)	Density (g/cm³)
MWNT	1200	-150	2.6
SWNT	1054	75	1.3
SWNT bundle	563	-150	1.3
Graphite (in-plane)	350	2.5	2.6
Steel	208	0.4	7.8

The elastic response of a nanotube to deformation is also very remarkable. Most hard materials fail with a strain of 1% or less due to propagation of dislocations and defects. Both theory and experiment show that CNTs can sustain up to 15% tensile strain before fracture. Thus the tensile strength of individual nanotube can be as high as 150 GPa, assuming 1 TPa for Young's modulus. Such a high strain is attributed to an elastic buckling through which high stress is released. Elastic buckling also exists in twisting and bending deformation of nanotubes. All elastic deformation including tensile (stretching and compression), twisting, and bending in a nanotube is nonlinear, featured by elastic buckling up to ~15% or even higher strain. This is another unique property of nanotubes, and such a high elastic strain for several deformation modes is originated from sp^2 rehybridization in nanotubes through which the high strain gets released.

However, sp^2 rehybridization will lead to change in electronic properties of a nanotube. A position vector in a deformed nanotube or graphite sheet can be written as $r = r_o + \Delta r$

where r can be deformed lattice vector a or chiral vector C. Using a similar approach to deriving electronic properties of a nanotube from graphite, the following relations are obtained:

$$E_g = E_{go} + s_{gn}(2p+1)3\gamma(1+v)(\cos 3\theta)\epsilon_l + (\sin 3\theta)\epsilon_r$$

In this relation, E_{go} is zero strain band gap, θ is nanotube chiral angle, ε_l and ε_r are tensile and torsion strain, respectively; and v is Poisson's ratio. Parameter p is defined by (n − m) = 3q + p such that p = 0 for metallic tube; p = 1 for type I semiconductor tube, for example, (10, 0); and p = −1 for type II semiconductor tube, for example, (8, 0). Thus, function $s_{gn}(2p+1)$ = 1, 1 and −1, respectively, for these three types of tubes. Equation $E_g = E_{go} + s_{gn}(2p+1)3\gamma(1+v)(\cos 3\theta)\epsilon_l + (\sin 3\theta)\epsilon_r$ predicts that all chiral or asymmetric tubes (0 < θ < 30°) will experience change in electronic properties for either tensile or torsional strain whereas symmetric armchair or zigzag tubes may or may not change their electronic properties. In asymmetric tubes, either strain will cause asymmetric σ - Π rehybridization and therefore change in electronic properties.

Magnetic Properties

The magnetic properties are studied with electron spin resonance (ESR), which is very important in understanding electronic properties, for example, for graphite and conjugated materials. Once again, there is a large discrepancy from different experimental measurements, especially in transport properties, because of sample quality and alignment whereas qualitatively they agree with theoretical calculations. Magnetic properties such as anisotropic g-factor and susceptibility of nanotubes are expected to be similar to those for graphite while some unusual properties may exist for nanotubes. Indeed, it is found from ESR that the average observed g-value of 2.012 and spin susceptibility of $7 \cdot 10^{-9}$ emu/g in MWNTs are only slightly lower than 2.018 and $2 \cdot 10^{-8}$ emu/g in graphite. Some interesting properties are also found from ESR studies of Pauli behavior. For example, aligned MWNTs are metallic or semi metallic. The measured susceptibility gives the density of state at the Fermi level of $1.5 \cdot 10^{-3}$ states/eV/atom, also comparable with that for in-plane graphite. The carrier concentration is about 10^{19} cm^{-3}, as compared with an upper limit of 10^{19} cm^{-3} from Hall measurement. However, similar observations have not been made for SWNTs and bundles. The possible reason is sample alignment difficulties and strong electron correlation, which may block non-conduction ESR signal.

Electrical Properties

The sp^2 bonds between carbon atoms results in conducting nature of carbon nanotubes. They can also withstand strong electric currents because of the strong nature of bonds. Single walled nanotubes can route electrical signals at speeds up to

10 GHz when used as interconnects on semi-conducting devices. Their electronic properties can be manipulated by application of external magnetic field, mechanical force etc.

Thermal Properties

CNTs can exhibit superconductivity below 20 K (approximately −253 °C) due to the strong in-plane C– C bonds of graphene. The strong C-C bond provides the exceptional strength and stiffness against axial strains. Moreover, the larger inter-plane and zero in-plane thermal expansion of SWNTs results in high flexibility against non-axial strains.

Due to their high thermal conductivity and large in-plane expansion, CNTs exhibit exciting prospects in nanoscale molecular electronics, sensing and actuating devices, reinforcing additive fibers in functional composite materials, etc. Recent experimental measurements suggest that the CNT embedded matrices are stronger in comparison to bare polymer matrices. Therefore, it is expected that the nanotube may also significantly improve the thermo-mechanical and the thermal properties of the composite materials.

Applications of CNT

The functionalization of CNTs makes them useful in a range of different applications. Their structure means that the tubes have an inner and an outer core which can both be modified by different functional groups. Thus the CNTs can be designed for very specific purposes. In the area of biomedicine, the applications of CNTs are investigated in especially four main fields: drug delivery, biomedical imaging, biosensors and scaffolds in tissue engineering.

Gene Therapy

CNTs can deliver a large amount of therapeutic agents, including DNA and RNA, to the target disease sites, Gene therapy and RNA have presented a great potential for antitumor treatment. The wire shaped structure (with a diameter matching that of DNA/siRNA) and their remarkable flexibility, CNTs can influence the conformational structure and the transient conformational changes of DNA RNA, which can further enhance the therapeutic effects of DNA is RNA. The treatment of a human lung carcinoma model in vivo using siRNA sequences, which led to cytotoxicity and cell death using amino-functionalized multi-walled carbon nanotubes (MWNT-NH_3^+). This is believed to activate biologically in vivo by triggering an apoptotic cascade that leads to extensive necrosis of the human tumor mass followed by a concomitant prolongation of survival of human lung tumor-bearing animals.

Composite Materials

Carbon nanotubes possess superior mechanical properties, and because of this fact the development of numerous structures such as smart clothes, sports gear, combat jackets, space elevators, etc., is expected to benefit tremendously from the use of CNTs. However, further improvement in the practical tensile strength of carbon nanotubes is necessary in the performance of space elevators; for this application, the CNT technology needs further advancements.

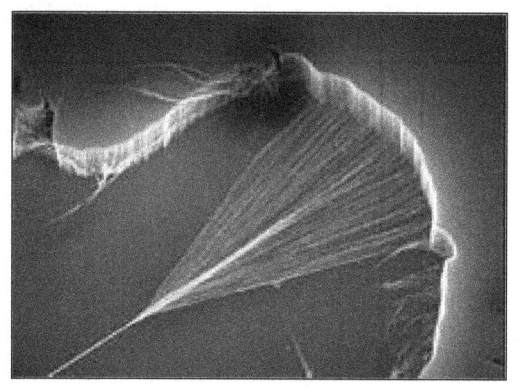

Figure: Spinning a yarn from CNTs.

Owing to their excellent mechanical properties, CNTs are expected to be behaving as building blocks for composite materials. Due to high strength, CNTs are being envisaged as promising materials to produce stab proof and bullet proof cloths. CNTs are supposed to prevent bullet from penetrating the body. However, the kinetic energy of the bullet may cause internal bleeding and some damage to bones. Figure above shows a yarn being spun from vertically aligned CNTs. The mechanical strength of these yarns is found to be very high.

Biomedical Imaging

Besides having unique electrical and mechanical properties CNTs also have optical properties that are very useful in applications such as biomedical imaging. SWNTs have strong optical absorption from ultraviolet (UV) to near infra-red (NIR) regions and are useful in a range of different imaging techniques. These include photoacoustic imaging, Raman imaging, fluorescence imaging, and with functionalization of the CNTs also positron emission tomography (PET) imaging and magnetic resonance (MR) imaging.

The CNTs were functionalized with a specific receptor for internalization into a specific cell type thus imaging these cells with very low autofluorescence background. In an in vivo study the biodistribution of SWNTs in live drosophila larvae was monitored by fluorescence imaging. In photoacoustic imaging deeper tissue penetration can be achieved compared to most other optical imaging techniques. The technique makes use of certain light absorbing molecules (for example CNTs) that converts laser pulses delivered into the biological tissue to heat. Thereby transient thermoelastic expansion is induced giving rise to wideband ultrasonic emission which can then be detected by an ultrasonic microphone. With their high optical absorption in the NIR range. SWNTs make a useful contrast agent in this kind of biomedical imaging.

Water Treatment

Carbon nanotubes are known to demonstrate strong adsorption affinities towards a wide range of aromatic and aliphatic contaminants in water, mainly because of their large and hydrophobic surfaces. They also show similar adsorption capacities when used as activated carbons in the presence of natural organic matter. Consequently, they are regarded as potential adsorbents for removal of contaminant in water and wastewater treatment systems.

Furthermore, membranes made from CNT arrays can be used as switchable molecular sieves, wherein their sieving and permeation features can be dynamically controlled by either pore size distribution (passive control) or external electrostatic fields (active control).

Batteries

Exciting electronic properties of carbon nanotubes (CNTs) have great potential for battery applications. CNTs are attracting interest as novel anode materials in the most popular lithium ion batteries (LIBs). Anodes in LIBs require high reversible capacity at a potential close to the lithium metal, and a moderate irreversible capacity. CNTs as anode materials have demonstrated these capabilities by enhancing the electrical conductivity as well as improving mechanical robustness of the electrode. This leads to enhanced rate capability and cyclic life of the battery.

Blood Cancer

Leukemia is a cancer that begins in the bone marrow (the soft inner part of some bones), but in most cases, moves into the blood. It can then spread to other parts of the body, such as organs and tissues. Acute lymphoblastic leukemia (ALL), one of the four main types of leukemia, is a slow-growing blood cancer that starts in bone marrow cells called lymphocytes or white blood cells. Once these white blood cells are affected by leukemia, they do not go through their normal process of maturing.

An intensified targeted delivery of daunorubicin (Dau) to acute lymphoblastic leukemia was achieved by Taghdisi et al., they developed a tertiary complex of Sgc8c aptamer (this aptamer targets leukemia biomarker protein tyrosine kinase-7) daunorubicin, and SWCNT named as Dau-aptamer SWCNTs. Flow cytometric analysis viewed that the tertiary complex was internalized effectively into human T cell leukemia cell (MOLT-4 cells) but not to U266 myeloma cells. Release of Dau-loaded nanotubes was pH-dependent. In a slightly acidic solution of pH 5.5, Dau was released from complex in 72 h at

37 °C, whilst Dau-aptamer-SWNTs tertiary complex was pretty stable after the same incubation at pH 7.4.

Mixtures

In the area of carbon nanotubes based composites; multi-walled carbon nanotubes (or MWCNTs) were the first to be exploited as electrically conducting fillers. MWCNTs were added to metal at high concentrations (~84 wt%). CNTs can also be used as fillers in insulating matrices. For example, a mere 10 % by weight addition of MWCNTs in polymer matrix increased the electrical conductivity up to 10,000S/m. CNTs based plastics are widely used in automotive industries for electrostatic painting of mirror housings, fuel lines as well as filters which dissipate electrostatic charge. Other uses of CNTs are in electromagnetic interference shielding.

The use of CNTs in the blades of wind turbines and hulls in boats for maritime security is based on the enhanced strength of the fiber composites using CNTs. CNTs can be made into stronger carbon finers having diameter of the order of few microns. In addition, the arrangement of carbon with pyrolyzed fibers is also affected by the presence of CNTs. Besides, the composites of concrete with CNTs enhance the tensile properties as well as crack performance of the resulting composites.

Textiles

Originally, CNTs' applications in textiles for enhancing the physical and mechanical behaviors of the fibers only included spinning the fibers. However, recent investigations have proposed coating of the textile fibres with CNTs. The findings of these experiments can be summarized as follows:

- Randomly oriented double-walled carbon nanotubes (DWCNTs) were coated with a thin film of polymer compounds to prepare DWCNTs/polymer composites. Afterwards, years were made from these composites by twisting and stretching ribbons out of it.

- These fibres were developed by Cambridge University and licensed for commercial production of body armors, e.g., combat jackets from them.

- Synthetic muscles have been prepared by CNT/polymer composites. They produce high contraction and extension ratio for a given amount of electric current.

References

- Carbon-nanotubes-Types-methods-of-preparation-and-applications-303994564: researchgate.net, Retrieved 18, July 2020

- Semiconductor-Materials-and-Devices, CMOS-technology: inflibnet.ac.in, Retrieved 04, April 2020

7

Consequences of Nanotechnology

There are several impacts and benefits of nanotechnology such as water purification systems, nanomedicine, improved manufacturing methods, better food production methods, etc. This chapter covers societal impact of nanotechnology, nanotoxicology, nanomaterials pollution and regulation of nanotechnology. The topics elaborated in this chapter will help in gaining a better perspective about the impacts of nanotechnology.

Impact of Nanotechnology

The impact of nanotechnology extends from its medical, ethical, mental, legal and environmental applications, to fields such as engineering, biology, chemistry, computing, materials science, and communications.

Major benefits of nanotechnology include improved manufacturing methods, water purification systems, energy systems, physical enhancement, nanomedicine, better food production methods, nutrition and large-scale infrastructure auto-fabrication. Nanotechnology's reduced size may allow for automation of tasks which were previously inaccessible due to physical restrictions, which in turn may reduce labor, land, or maintenance requirements placed on humans.

Potential risks include environmental, health, and safety issues; transitional effects such as displacement of traditional industries as the products of nanotechnology become dominant, which are of concern to privacy rights advocates. These may be particularly important if potential negative effects of nanoparticles are overlooked.

Whether nanotechnology merits special government regulation is a controversial issue. Regulatory bodies such as the United States Environmental Protection Agency and the Health and Consumer Protection Directorate of the European Commission have started dealing with the potential risks of nanoparticles. The organic food sector has been the first to act with the regulated exclusion of engineered nanoparticles from certified organic produce, firstly in Australia and the UK, and more recently in Canada, as well as for all food certified to Demeter International standards.

Overview

The presence of nanomaterials (materials that contain nanoparticles) is not in itself a threat. It is only certain aspects that can make them risky, in particular their mobility and their increased reactivity. Only if certain properties of certain nanoparticles were harmful to living beings or the environment would we be faced with a genuine hazard. In this case it can be called nanopollution.

In addressing the health and environmental impact of nanomaterials we need to differentiate between two types of nanostructures: (1) Nanocomposites, nanostructured surfaces and nanocomponents (electronic, optical, sensors etc.), where nanoscale particles are incorporated into a substance, material or device ("fixed" nano-particles); and (2) "free" nanoparticles, where at some stage in production or use individual nanoparticles of a substance are present. These free nanoparticles could be nanoscale species of elements, or simple compounds, but also complex compounds where for instance a nanoparticle of a particular element is coated with another substance ("coated" nanoparticle or "core-shell" nanoparticle).

There seems to be consensus that, although one should be aware of materials containing fixed nanoparticles, the immediate concern is with free nanoparticles.

Nanoparticles are very different from their everyday counterparts, so their adverse effects cannot be derived from the known toxicity of the macro-sized material. This poses significant issues for addressing the health and environmental impact of free nanoparticles.

To complicate things further, in talking about nanoparticles it is important that a powder or liquid containing nanoparticles almost never be monodisperse, but contain instead a range of particle sizes. This complicates the experimental analysis as larger nanoparticles might have different properties from smaller ones. Also, nanoparticles show a tendency to aggregate, and such aggregates often behave differently from individual nanoparticles.

Health Impact

The health impacts of nanotechnology are the possible effects that the use of nanotechnological materials and devices will have on human health. As nanotechnology is an emerging field, there is great debate regarding to what extent nanotechnology will benefit or pose risks for human health. Nanotechnology's health impacts can be split into two aspects: the potential for nanotechnological innovations to have medical applications to cure disease, and the potential health hazards posed by exposure to nanomaterials.

Medical Applications

Nanomedicine is the medical application of nanotechnology. The approaches to nanomedicine range from the medical use of nanomaterials, to nanoelectronic biosensors, and even possible future applications of molecular nanotechnology. Nanomedicine seeks to deliver a valuable set of research tools and clinically helpful devices in the near future. The National Nanotechnology Initiative expects new commercial applications in the pharmaceutical industry that may include advanced drug delivery systems, new therapies, and in vivo imaging. Neuro-electronic interfaces and other nanoelectronics-based sensors are another active goal of research. Further down the line, the speculative field of molecular nanotechnology believes that cell repair machines could revolutionize medicine and the medical field.

Nanomedicine seeks to deliver a set of research tools and clinical devices in the near future. The National Nanotechnology Initiative expects new commercial applications in the pharmaceutical industry that may include advanced drug delivery systems, new therapies, and *in vivo* imaging. Neuro-electronic interfaces and other nanoelectronics-based sensors are another active goal of research. Further down the line, the speculative field of molecular nanotechnology believes that cell repair machines could revolutionize medicine and the medical field.

Nanomedicine research is directly funded, with the US National Institutes of Health in 2005 funding a five-year plan to set up four nanomedicine centers. In April 2006, the journal Nature Materials estimated that 130 nanotech-based drugs and delivery systems were being developed worldwide. Nanomedicine is a large industry, with nanomedicine sales reaching $6.8 billion in 2004. With over 200 companies and 38 products worldwide, a minimum of $3.8 billion in nanotechnology R&D is being invested every year. As the nanomedicine industry continues to grow, it is expected to have a significant impact on the economy.

Health Hazards

Nanotoxicology is the field which studies potential health risks of nanomaterials. The extremely small size of nanomaterials means that they are much more readily taken up by the human body than larger sized particles. How these nanoparticles behave inside the organism is one of the significant issues that needs to be resolved. The behavior of nanoparticles is a function of their size, shape and surface reactivity with the surrounding tissue. Apart from what happens if non-degradable or slowly degradable nanoparticles accumulate in organs, another concern is their potential interaction with biological processes inside the body: because of their large surface, nanoparticles on exposure to tissue and fluids will immediately adsorb onto their surface some of the macromolecules they encounter. The large number of variables influencing toxicity means that it is difficult to generalise about health risks associated with exposure to nanomaterials – each new nanomaterial must be assessed individually and all material properties must be taken into account. Health and environmental issues combine in the workplace of companies engaged in producing or using nanomaterials and in the laboratories engaged in nanoscience and nanotechnology research. It is safe to say that current workplace exposure standards for dusts cannot be applied directly to nanoparticle dusts.

The extremely small size of nanomaterials also means that they are much more readily taken up by the human body than larger sized particles. How these nanoparticles behave inside the body is one of the issues that needs to be resolved. The behavior of nanoparticles is a function of their size, shape and surface reactivity with the surrounding tissue. They could cause overload on phagocytes, cells that ingest and destroy foreign matter, thereby triggering stress reactions that lead to inflammation and weaken the body's defense against other pathogens. Apart from what happens if non-degradable or slowly degradable nanoparticles accumulate in organs, another concern is their potential interaction with biological processes inside the body: because of their large surface, nanoparticles on exposure to tissue and fluids will immediately adsorb onto their surface some of the macromolecules they encounter. This may, for instance, affect the regulatory mechanisms of enzymes and other proteins.

The National Institute for Occupational Safety and Health has conducted initial research on how nanoparticles interact with the body's systems and how workers might be exposed to nano-sized particles in the manufacturing or industrial use of nanomaterials. NIOSH currently offers interim guidelines for working with nanomaterials consistent with the best scientific knowledge. At The National Personal Protective Technology Laboratory of NIOSH, studies investigating the filter penetration of nanoparticles on NIOSH-certified and EU marked respirators, as well as non-certified dust masks have been conducted. These studies found that the most penetrating particle size range was between 30 and 100 nanometers, and leak size was the largest factor in the number of nanoparticles found inside the respirators of the test dummies.

Other properties of nanomaterials that influence toxicity include: chemical composition, shape, surface structure, surface charge, aggregation and solubility, and the presence or absence of functional groups of other chemicals. The large number of variables influencing toxicity means that it is difficult to generalise about health risks associated with exposure to nanomaterials – each new nanomaterial must be assessed individually and all material properties must be taken into account.

Literature reviews have been showing that release of engineered nanoparticles and incurred personal exposure can happen during different work activities. The situation alerts regulatory bodies to necessitate prevention strategies and regulations at nanotechnology workplaces.

Environmental Impact

The environmental impact of nanotechnology is the possible effects that the use of nanotechnological materials and devices will have on the environment. As nanotechnology is an emerging field, there is great debate regarding to what extent industrial and commercial use of nanomaterials will affect organisms and ecosystems.

Nanotechnology's environmental impact can be split into two aspects: the potential for nanotechnological innovations to help improve the environment, and the possibly novel type of pollution that nanotechnological materials might cause if released into the environment.

Environmental Applications

Green nanotechnology refers to the use of nanotechnology to enhance the environmental sustainability of processes producing negative externalities. It also refers to the use of the products of nanotechnology to enhance sustainability. It includes making green nano-products and using nano-products in support of sustainability. Green nanotechnology has been described as the development of clean technologies, "to minimize potential environmental and human health risks associated with the manufacture and use of nanotechnology products, and to encourage replacement of existing products with new nano-products that are more environmentally friendly throughout their lifecycle."

Green nanotechnology has two goals: producing nanomaterials and products without harming the environment or human health, and producing nano-products that provide solutions to environmental problems. It uses existing principles of green chemistry and green engineering to make nanomaterials and nano-products without toxic ingredients, at low temperatures using less energy and renewable inputs wherever possible, and using lifecycle thinking in all design and engineering stages.

Pollution

Nanopollution is a generic name for all waste generated by nanodevices or during the nanomaterials manufacturing process. Nanowaste is mainly the group of particles that are released into the environment, or the particles that are thrown away when still on their products.

Societal Impact

Beyond the toxicity risks to human health and the environment which are associated with first-generation nanomaterials, nanotechnology has broader societal impact and poses broader social challenges. Social scientists have suggested that nanotechnology's social issues should be understood and assessed not simply as "downstream" risks or impacts. Rather, the challenges should be factored into "upstream" research and decision-making in order to ensure technology development that meets social objectives.

Many social scientists and organizations in civil society suggest that technology assessment and governance should also involve public participation.

Over 800 nano-related patents were granted in 2003, with numbers increasing to nearly 19,000 internationally by 2012. Corporations are already taking out broad-ranging patents on nanoscale discoveries and inventions. For example, two corporations, NEC and IBM, hold the basic patents on carbon nanotubes, one of the current cornerstones of nanotechnology. Carbon nanotubes have a wide range of uses, and look set to become crucial to several industries from electronics and computers, to strengthened materials to drug delivery and diagnostics. Carbon nanotubes are poised to become a major traded commodity with the potential to replace major conventional raw materials.

Nanotechnologies may provide new solutions for the millions of people in developing countries who lack access to basic services, such as safe water, reliable energy, health care, and education. The 2004 UN Task Force on Science, Technology and Innovation noted that some of the advantages of nanotechnology include production using little labor, land, or maintenance, high productivity, low cost, and modest requirements for materials and energy. However, concerns are frequently raised that the claimed benefits of nanotechnology will not be evenly distributed, and that any benefits (including technical and/or economic) associated with nanotechnology will only reach affluent nations.

Longer-term concerns center on the impact that new technologies will have for society at large, and whether these could possibly lead to either a post-scarcity economy, or alternatively exacerbate the wealth gap between developed and developing nations. The effects of nanotechnology on the society as a whole, on human health and the environment, on trade, on security, on food systems and even on the definition of "human", have not been characterized or politicized.

Regulation

Significant debate exists relating to the question of whether nanotechnology or nanotechnology-based products merit special government regulation. This debate is related to the circumstances in which it is necessary and appropriate to assess new substances prior to their release into the market, community and environment.

Regulatory bodies such as the United States Environmental Protection Agency and the Food and Drug Administration in the U.S. or the Health & Consumer Protection Directorate of the European Commission have started dealing with the potential risks posed by nanoparticles. So far, neither engineered nanoparticles nor the products and materials that contain them are subject to any special regulation regarding production, handling or labelling. The Material Safety Data Sheet that must be issued for some materials often does not differentiate between bulk and nanoscale size of the material in question and even when it does these MSDS are advisory only.

Limited nanotechnology labeling and regulation may exacerbate potential human and environmental health and safety issues associated with nanotechnology. It has been argued that the development of comprehensive regulation of nanotechnology will be vital to ensure that the potential risks associated with the research and commercial application of nanotechnology do not overshadow its potential benefits. Regulation may also be required to meet community expectations about responsible development of nanotechnology, as well as ensuring that public interests are included in shaping the development of nanotechnology.

In "The Consumer Product Safety Commission and Nanotechnology," E. Marla Felcher suggests that the Consumer Product Safety Commission, which is charged with protecting the public against unreasonable risks of injury or death associated with consumer products, is ill-equipped to oversee the safety of complex, high-tech products made using nanotechnology.

Societal Impact of Nanotechnology

The societal impact of nanotechnology are the potential benefits and challenges that the introduction of novel nanotechnological devices and materials may hold for society and human interaction. The term is sometimes expanded to also include nanotechnology's health and environmental impact, but this article will only consider the social and political impact of nanotechnology.

As nanotechnology is an emerging field and most of its applications are still speculative, there is much debate about what positive and negative effects that nanotechnology might have.

Overview

Beyond the toxicity risks to human health and the environment which are associated with first-generation nanomaterials, nanotechnology has broader societal implications and poses broader social challenges. Social scientists have suggested that nanotechnology's social issues should be understood and assessed not simply as "downstream" risks or impacts. Rather, the challenges should be factored into "upstream" research and decision making in order to ensure technology development that meets social objectives.

Many social scientists and organizations in civil society suggest that technology assessment and governance should also involve public participation

Some observers suggest that nanotechnology will build incrementally, as did the 18-19th century industrial revolution, until it gathers pace to drive a nanotechnological revolution that will radically reshape our economies, our labor markets, international trade, international relations, social structures, civil liberties, our relationship with the natural world and even what we understand to be human. Others suggest that it may be more accurate to describe change driven by nanotechnology as a "technological tsunami". Just like a tsunami, analysts warn that rapid nanotechnology-driven change will necessarily have profound disruptive impacts. As the APEC Center for Technology Foresight observes:

If nanotechnology is going to revolutionize manufacturing, health care, energy supply, communications and probably defense, then it will transform labour and the workplace, the medical system, the transportation and power infrastructures and the military. None of these latter will be changed without significant social disruption.

Those concerned with the negative impact of nanotechnology suggest that it will simply exacerbate problems stemming from existing socio-economic inequity and unequal distributions of power, creating greater inequities between rich and poor through an inevitable nano-divide (the gap between those who control the new nanotechnologies and those whose products, services or labour are displaced by them). Analysts suggest the possibility that nanotechnology has the potential to destabilize international relations through a nano arms race and the increased potential for bioweaponry; thus, providing the tools for ubiquitous surveillance with significant implications for civil liberties. Also, many critics believe it might break down the barriers between life and non-life through nanobiotechnology, redefining even what it means to be human.

Nanoethicists posit that such a transformative technology could exacerbate the divisions of rich and poor – the so-called "nano divide." However nanotechnology makes the production of technology, e.g. computers, cellular phones, health technology etcetera, cheaper and therefore accessible to the poor.

In fact, many of the most enthusiastic proponents of nanotechnology, such as transhumanists, see the nascent science as a mechanism to changing human nature itself – going beyond curing disease and enhancing human characteristics. Discussions on nanoethics have been hosted by the federal government, especially in the context of "converging technologies" – a catch-phrase used to refer to nano, biotech, information technology, and cognitive science.

Possible Military Applications

Possible military applications of nanotechnology have been suggested in the fields of soldier enhancement () and chemical weapons amongst others. However, more socially disruptive weapon systems are to be expected from molecular manufacturing, a potential future form of nanotechnology that would make it possible to build complex structures at atomic precision. Molecular manufacturing requires significant advances in nanotechnology, but its supporters posit that once achieved it could produce highly advanced products at low costs and in large quantities in nanofactories weighing a kilogram or more. If nanofactories gain the ability to produce other nanofactories production may only be limited by relatively abundant factors such as input materials, energy and software.

Molecular manufacturing might be used to cheaply produce, among many other products, highly advanced, durable weapons. Being equipped with compact computers and motors these might be increasingly autonomous and have a large range of capabilities.

According to Chris Phoenix and Mike Treder from the Center for Responsible Nanotechnology as well as Anders Sandberg from the Future of Humanity Institute the military uses of molecular manufacturing are the applications of nanotechnology that pose the most significant global catastrophic risk. Several nanotechnology researchers state that the bulk of risk from nanotechnology comes from the potential to lead to war, arms races and destructive global government. Several reasons have been suggested why the availability of nanotech weaponry may with significant likelihood lead to unstable arms races (compared to e.g. nuclear arms races): (1) A large number of players may be tempted to enter the race since the threshold for doing so is low; (2) the ability to make weapons with molecular manufacturing might be cheap and easy to hide; (3) therefore lack of insight into the other parties' capabilities can tempt players to arm out of caution or to launch preemptive strikes; (4) molecular manufacturing may reduce dependency on international trade, a potential peace-promoting factor; (5) wars of aggression may pose a smaller economic threat to the aggressor since manufacturing is cheap and humans may not be needed on the battlefield.

Self-regulation by all state and non-state actors has been called hard to achieve, so measures to mitigate war-related risks have mainly been proposed in the area of international cooperation. International infrastructure may be expanded giving more sovereignty to the international level. This could help coordinate efforts for arms control. Some have put forth that international institutions dedicated specifically to nanotechnology (perhaps analogously to the International Atomic Energy Agency IAEA) or general arms control may also be designed. One may also jointly make differential technological progress on defensive technologies. The Center for Responsible Nanotechnology also suggest some technical restrictions. Improved transparency regarding technological capabilities may be another important facilitator for arms-control.

Intellectual Property Issues

On the structural level, critics of nanotechnology point to a new world of ownership and corporate control opened up by nanotechnology. The claim is that, just as biotechnology's ability to manipulate genes went hand in hand with the patenting of life, so too nanotechnology's ability to manipulate molecules has led to the patenting of matter. The last few years has seen a gold rush to claim patents at the nanoscale. Academics have warned that the resultant patent thicket is harming progress in the technology and have argued in the top journal *Nature* that there should be a moratorium on patents on "building block" nanotechnologies. Over 800 nano-related patents were granted in 2003, and the numbers are increasing year to year. Corporations are already taking out broad-ranging patents on nanoscale discoveries and inventions. For example, two corporations, NEC and IBM, hold the basic patents on carbon nanotubes, one of the current cornerstones of nanotechnology. Carbon nanotubes have a wide range of uses, and look set to become crucial to several industries from electronics and computers, to strengthened materials to drug delivery and diagnostics. Carbon nanotubes are poised to become a major traded commodity with the potential to replace major conventional raw materials. However, as their use expands, anyone seeking to (legally) manufacture or sell carbon nanotubes, no matter what the application, must first buy a license from NEC or IBM.

The United States' essential facilities doctrine may be of importance as well as other anti-trust laws.

Potential Benefits and Risks for Developing Countries

Nanotechnologies may provide new solutions for the millions of people in developing countries who lack access to basic services, such as safe water, reliable energy, health care, and education. The United Nations has set Millennium Development Goals for meeting these needs. The 2004 UN Task Force on Science, Technology and Innovation noted that some of the advantages of nanotechnology include production using little labor, land, or maintenance, high productivity, low cost, and modest requirements for materials and energy.

Many developing countries, for example Costa Rica, Chile, Bangladesh, Thailand, and Malaysia, are investing considerable resources in research and development of nanotechnologies. Emerging economies such as Brazil, China, India and South Africa are spending millions of US dollars annually on R&D, and are rapidly increasing their scientific output as demonstrated by their increasing numbers of publications in peer-reviewed scientific publications.

Potential opportunities of nanotechnologies to help address critical international development priorities include improved water purification systems, energy systems, medicine and pharmaceuticals, food production and nutrition, and information and communications technologies. Nanotechnologies are already incorporated in products that are on the market. Other nanotechnologies are still in the research phase, while others are concepts that are years or decades away from development.

Applying nanotechnologies in developing countries raises similar questions about the environmental, health, and societal risks described in the previous section. Additional challenges have been raised regarding the linkages between nanotechnology and development.

Protection of the environment, human health and worker safety in developing countries often suffers from a combination of factors that can include but are not limited to lack of robust environmental, human health, and worker safety regulations; poorly or unenforced regulation which is linked to a lack of physical (e.g., equipment) and human capacity (i.e., properly trained regulatory staff). Often, these nations require assistance, particularly financial assistance, to develop the scientific and institutional capacity to adequately assess and manage risks, including the necessary infrastructure such as laboratories and technology for detection.

Very little is known about the risks and broader impacts of nanotechnology. At a time of great uncertainty over the impacts of nanotechnology it will be challenging for governments, companies, civil society organizations, and the general public in developing countries, as in developed countries, to make decisions about the governance of nanotechnology.

Companies, and to a lesser extent governments and universities, are receiving patents on nanotechnology. The rapid increase in patenting of nanotechnology is illustrated by the fact that in the US, there were 500 nanotechnology patent applications in 1998 and 1,300 in 2000. Some patents are very broadly defined, which has raised concern among some groups that the rush to patent could slow innovation and drive up costs of products, thus reducing the potential for innovations that could benefit low income populations in developing countries.

There is a clear link between commodities and poverty. Many least developed countries are dependent on a few commodities for employment, government revenue, and export earnings. Many applications of nanotechnology are being developed that could impact global demand for specific commodities. For instance, certain nanoscale materials could enhance the strength and durability of rubber, which might eventually lead to a decrease in demand for natural rubber. Other nanotechnology applications may result in increases in demand for certain commodities. For example, demand for titanium may increase as a result of new uses for nanoscale titanium oxides, such as titanium dioxide nanotubes that can be used to produce and store hydrogen for use as fuel. Various organizations have called for international dialogue on mechanisms that will allow developing countries to anticipate and proactively adjust to these changes.

In 2003, Meridian Institute began the Global Dialogue on Nanotechnology and the Poor: Opportunities and Risks (GDNP) to raise awareness of the opportunities and risks of nanotechnology for developing countries, close the gaps within and between sectors of society to catalyze actions that address specific opportunities and risks of nanotechnology for developing countries, and identify ways that science and technology can play an appropriate role in the development process. The GDNP has released several publicly accessible papers on nanotechnology and development, including "Nanotechnology and the Poor: Opportunities and Risks - Closing the Gaps Within and Between Sectors of Society"; "Nanotechnology, Water, and Development"; and "Overview and Comparison of Conventional and Nano-Based Water Treatment Technologies".

Social Justice and Civil Liberties

Concerns are frequently raised that the claimed benefits of nanotechnology will not be evenly distributed, and that any benefits (including technical and/or economic) associated with nanotechnology will only reach affluent nations. The majority of nanotechnology research and development - and patents for nanomaterials and products - is concentrated in developed countries (including the United States, Japan, Germany, Canada and France). In addition, most patents related to nanotechnology are concentrated amongst few multinational corporations, including IBM, Micron Technologies, Advanced Micro Devices and Intel. This has led to fears that it will be unlikely that developing countries will have access to the infrastructure, funding and human resources required to support nanotechnology research and development, and that this is likely to exacerbate such inequalities.

Producers in developing countries could also be disadvantaged by the replacement of natural products (including rubber, cotton, coffee and tea) by developments in nanotechnology. These natural products are important export crops for developing countries, and many farmers' livelihoods depend on them. It has been argued that their substitution with industrial nano-products could negatively impact the economies of developing countries, that have traditionally relied on these export crops.

It is proposed that nanotechnology can only be effective in alleviating poverty and aid development "when adapted to social, cultural and local institutional contexts, and chosen and designed with the active participation by citizens right from the commencement point" (Invernizzi et al. 2008, p. 132).

Effects on Laborers

Ray Kurzweil has speculated in *The Singularity is Near* that people who work in unskilled labor jobs for a livelihood may become the first human workers to be displaced by the constant use of nanotechnology in the workplace, noting that layoffs often affect the jobs based around the lowest technology level before attacking jobs with the highest technology level possible. It has been noted that every major economic era has stimulated a global revolution both in the kinds of jobs that are available to people and the kind of training they need to achieve these jobs, and there is concern that the world's educational systems have lagged behind in preparing students for the "Nanotech Age".

It has also been speculated that nanotechnology may give rise to nanofactories which may have superior capabilities to conventional factories due to their small carbon and physical footprint on the global and regional environment. The miniaturization and transformation of the multi-acre conventional factory into the nanofactory may not interfere with their ability to deliver a high quality product; the product may be of even greater quality due to the lack of human errors in the production stages. Nanofactory systems may use precise atomic precisioning and contribute to making superior quality products that the "bulk chemistry" method used in 20th century and early 21st currently cannot produce. These advances might shift the computerized workforce in an even more complex direction, requiring skills in genetics, nanotechnology, and robotics.

Nanotoxicology

Nanotoxicology is the study of the toxicity of nanomaterials. Because of quantum size effects and large surface area to volume ratio, nanomaterials have unique properties compared with their larger counterparts.

Nanotoxicology is a branch of bionanoscience which deals with the study and application of toxicity of nanomaterials. Nanomaterials, even when made of inert elements like gold, become highly active at nanometer dimensions. Nanotoxicological studies are intended to determine whether and to what extent these properties may pose a threat to the environment and to human beings. For instance, Diesel nanoparticles have been found to damage the cardiovascular system in a mouse model.

Background

Nanotoxicology is a sub-specialty of particle toxicology. It addresses the toxicology of nanoparticles (particles <100 nm diameter) which appear to have toxicity effects that are unusual and not seen with larger particles. Nanoparticles can be divided into combustion-derived nanoparticles (like diesel soot), manufactured nanoparticles like carbon nanotubes and naturally occurring nanoparticles from volcanic eruptions, atmospheric chemistry etc. Typical nanoparticles that have been studied are titanium dioxide, alumina, zinc oxide, carbon black, and carbon nanotubes, and "nano-C_{60}". Nanoparticles have much larger surface area to unit mass ratios which in some cases may lead to greater pro-inflammatory effects (in, for example, lung tissue). In addition, some nanoparticles seem to be able to translocate from their site of deposition to distant sites such as the blood and the brain. This has resulted in a sea-change in how particle toxicology is viewed- instead of being confined to the lungs, nanoparticle toxicologists study the brain, blood, liver, skin and gut.

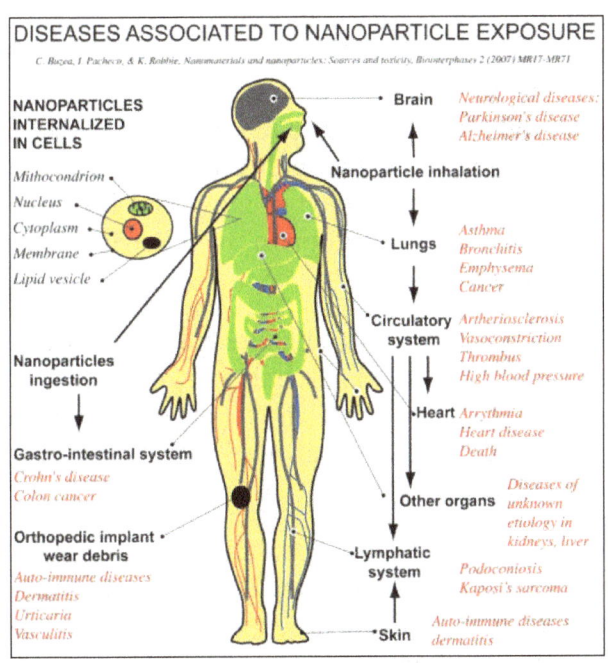

Pathways of exposure to nanoparticles and associated diseases as suggested by epidemiological, in vivo and in vitro studies.

Calls for tighter regulation of nanotechnology have arisen alongside a growing debate related to the human health and safety risks associated with nanotechnology. From a large-scale literature review, Yaobo Ding et al. found that release of airborne engineered nanoparticles and associated worker exposure from various production and handling activities at different workplaces are very probable. The Royal Society identifies the potential for nanoparticles to penetrate the skin, and recommends that the use of nanoparticles in cosmetics be conditional upon a favorable assessment by the relevant European Commission safety advisory committee. Andrew Maynard also reports that 'certain nanoparticles may move easily into sensitive lung tissues after inhalation, and cause damage that can lead to chronic breathing problems'.

Carbon nanotubes – characterized by their microscopic size and incredible tensile strength – are frequently likened to asbestos, due to their needle-like fiber shape. In a recent study that introduced carbon nanotubes into the abdominal cavity of mice, results demonstrated that long thin carbon nanotubes showed the same effects as long thin asbestos fibers, raising concerns that exposure to carbon nanotubes may lead to pleural abnormalities such as mesothelioma (cancer of the lining of the lungs caused by exposure to asbestos). Given these risks, effective and rigorous regulation has been called for to determine if, and under what circumstances, carbon nanotubes are manufactured, as well as ensuring their safe handling and disposal.

The Woodrow Wilson Centre's Project on Emerging Technologies conclude that there is insufficient funding for human health and safety research, and as a result there is currently limited understanding of the human health and safety risks associated with nanotechnology. While the US National Nanotechnology Initiative reports that around four percent (about $40 million) is dedicated to risk related research and development, the Woodrow Wilson Centre estimate that only around $11 million is actually directed towards risk related research. They argued in 2007 that it would be necessary to increase funding to a minimum of $50 million in the following two years so as to fill the gaps in knowledge in these areas.

The potential for workplace exposure was highlighted by the 2004 Royal Society report which recommended a review of existing regulations to assess and control workplace exposure to nanoparticles and nanotubes. The report expressed particular concern for the inhalation of large quantities of nanoparticles by workers involved in the manufacturing process.

Stakeholders concerned by the lack of a regulatory framework to assess and control risks associated with the release of nanoparticles and nanotubes have drawn parallels with bovine spongiform encephalopathy ('mad cow's disease'), thalidomide, genetically modified food, nuclear energy, reproductive technologies, biotechnology, and asbestosis. In light of such concerns, the Canadian based ETC Group have called for a moratorium on nano-related research until comprehensive regulatory frameworks are developed that will ensure workplace safety.

Reactive Oxygen Species

For some types of particles, the smaller they are, the greater their surface area to volume ratio and the higher their chemical reactivity and biological activity. The greater chemical reactivity of nanomaterials can result in increased production of reactive oxygen species (ROS), including free radicals. ROS production has been found in a diverse range of nanomaterials including carbon fullerenes, carbon nanotubes and nanoparticle metal oxides. ROS and free radical production is one of the primary mechanisms of nanoparticle toxicity; it may result in oxidative stress, inflammation, and consequent damage to proteins, membranes and DNA.

Biodistribution

The extremely small size of nanomaterials also means that they much more readily gain entry into the human body than larger sized particles. How these nanoparticles behave inside the body is still a major question that needs to be resolved. The behavior of nanoparticles is a function of their size, shape and surface reactivity with the surrounding tissue. In principle, a large number of particles could overload the body's phagocytes, cells that ingest and destroy foreign matter, thereby triggering stress reactions that lead to inflammation and weaken the body's defense against other pathogens. In addition to questions about what happens if non-degradable or slowly degradable nanoparticles accumulate in bodily organs, another concern is their potential interaction or interference with biological processes inside the body. Because of their large surface area, nanoparticles will, on exposure to tissue and fluids, immediately adsorb onto their surface some of the macromolecules they encounter. This may, for instance, affect the regulatory mechanisms of enzymes and other proteins.

Nanomaterials are able to cross biological membranes and access cells, tissues and organs that larger-sized particles normally cannot. Nanomaterials can gain access to the blood stream via inhalation or ingestion. At least some nanomaterials can penetrate the skin; even larger microparticles may penetrate skin when it is flexed. Broken skin is an ineffective particle barrier, suggesting that acne, eczema, shaving wounds or severe sunburn may accelerate skin uptake of nanomaterials. Then, once in the blood stream, nanomaterials can be transported around the body and be taken up by organs and tissues, including the brain, heart, liver, kidneys, spleen, bone marrow and nervous system. Nanomaterials have proved toxic to human tissue and cell cultures, resulting in increased oxidative stress, inflammatory cytokine production and cell death. Unlike larger particles, nanomaterials may be taken up by cell mitochondria and the cell nucleus. Studies demonstrate the potential for nanomaterials to cause DNA mutation and induce major structural damage to mitochondria, even resulting in cell death.

Nanotoxicity Studies

There is presently no authority to specifically regulate nanotech-based products. Scientific research has indicated the potential for some nanomaterials to be toxic to humans or the environment. In March 2004 tests conducted by environmental toxicologist Eva Oberdörster, Ph.D. working with Southern Methodist University in Texas, found extensive brain damage to fish exposed to fullerenes for a period of just 48 hours at a relatively moderate dose of 0.5 parts per million (commensurate with levels of other kinds of pollution found in bays). The fish also exhibited changed gene markers in their livers, indicating their entire physiology was affected. In a concurrent test, the fullerenes killed water fleas, an important link in the marine food chain. The extremely small size of fabricated nanomaterials also means that they are much more readily taken up by living tissue than presently known toxins. Nanoparticles can be inhaled, swallowed, absorbed through skin and deliberately or accidentally injected during medical procedures. They might be accidentally or inadvertently released from materials implanted into living tissue.

Researcher Shosaku Kashiwada of the National Institute for Environmental Studies in Tsukuba, Japan, in a more recent study, intended to further investigate the effects of nanoparticles on soft-bodied organisms. His study allowed him to explore the distribution of water-suspended fluorescent nanoparticles throughout the eggs and adult bodies of a species of fish, known as the see-through medaka (Oryzias latipes). See-through medaka were used because of their small size, wide temperature and salinity tolerances, and short generation time. Moreover, small fish like the see-through medaka have been popular test subjects for human diseases and organogenesis for other reasons as well, including their transparent embryos, rapid embryo development, and the functional equivalence of their organs and tissue material to that of mammals. Because the see-through medaka have transparent bodies, analyzing the deposition of fluorescent nanoparticles throughout the body is quite simple. For his study, Dr. Kashiwada evaluated four aspects of nanoparticle accumulation. These included the overall accumulation and the size-dependent accumulation of nanoparticles by medaka eggs, the effects of salinity on the aggregation of nanoparticles in solution and on their accumulation by medaka eggs, and the distribution of nanoparticles in the blood and organs of adult medaka. It was also noted that nanoparticles were in fact taken up into the bloodstream and deposited throughout the body. In the medaka eggs, there was a high accumulation of nanoparticles in the yolk; most often bioavailability was dependent on specific sizes of the particles. Adult samples of medaka had accumulated nanoparticles in the gills, intestine, brain, testis, liver, and bloodstream. One major result from this study was the fact that salinity may have a large influence on the bioavailibility and toxicity of nanoparticles to penetrate membranes and eventually kill the specimen.

As the use of nanomaterials increases worldwide, concerns for worker and user safety are mounting. To address such concerns, the Swedish Karolinska Institute conducted a study in which various nanoparticles were introduced to human lung epithelial cells. The results, released in 2008, showed that iron oxide nanoparticles caused little DNA damage and were non-toxic. Zinc oxide nanoparticles were slightly worse. Titanium dioxide caused only DNA damage. Carbon nanotubes caused DNA damage at low levels. Copper oxide was found to be the worst offender, and was the only nanomaterial identified by the researchers as a clear health risk. The latest toxicology studies on mice involving exposure to carbon nanotubes (CNT) showed a limited pulmonary inflammatory potential of MWCNT at levels corresponding to the average inhalable elemental carbon concentrations observed in U.S.-based CNT facilities. The study estimated that considerable years of exposure are necessary for significant pathology to occur.

No Fullerene Toxicity Reported

Nanoparticles can also be made of C_{60}, as is the case with almost any room temperature solid, and several groups have done this and studied toxicity of such particles. The results in the work of Oberdörster at Southern Methodist University, published in "Environmental Health Perspectives" in July 2004, in which questions were raised of potential cytotoxicity, has now been shown by several sources to be likely caused by the tetrahydrofuran used in preparing the 30 nm–100 nm particles of C_{60} used in the research. Isakovic, et al., 2006, who review this phenomenon, gives results showing that removal of THF from the C_{60} particles resulted in a loss of toxicity. Sayes, et al., 2007, also show that particles prepared as in Oberdorster caused no detectable inflammatory response when instilled intratracheally in rats after observation for 3 months, suggesting that even the particles prepared by Oberdorster do not exhibit markers of toxicity in mammalian models. This work used as a benchmark quartz particles, which did give an inflammatory response.

A comprehensive and recent review of work on fullerene toxicity is available in "Toxicity Studies of Fullerenes and Derivatives," a chapter from the book "Bio-applications of Nanoparticles". In this work, the authors review the work on fullerene toxicity beginning in the early 1990s to present, and conclude that the evidence gathered since the discovery of fullerenes overwhelmingly points to C_{60} being non-toxic. As is the case for toxicity profile with any chemical modification of a structural moiety, the authors suggest that individual molecules be assessed individually.

Toxicity of Metal Based Nanoparticles

Metal based nanoparticles (NPs) are a prominent class of NPs synthesized for their functions as semiconductors, electroluminescents, and thermoelectric materials. Biomedically, these antibacterial NPs have been utilized in drug delivery systems to access areas previously inaccessible to conventional medicine. With the recent increase in interest and development of nanotechnology, many studies have been performed to assess whether the unique characteristics of these NPs, namely their small surface area to volume ratio, might negatively impact the environment upon which they were introduced. Researchers have since found that many metal and metal oxide NPs have detrimental effects on the cells with which they come into contact including but not limited to DNA breakage and oxidation, mutations, reduced cell viability, warped morphology, induced apoptosis and necrosis, and decreased proliferation.

Cytotoxicity

A primary marker for the damaging effects of NPs has been cell viability as determined by state and exposed surface area of the cell membrane. Cells exposed to metallic NPs have, in the case of copper oxide, had up to 60% of their cells rendered unviable. When diluted, the positively charged metal ions often experience an electrostatic attraction to the cell membrane of nearby cells, covering the membrane and preventing it from permeating the necessary fuels and wastes. With less exposed membrane for transportation and communication, the cells are often rendered inactive.

NPs have been found to induce apoptosis in certain cells primarily due to the mitochondrial damage and oxidative stress brought on by the foreign NPs electrostatic reactions.

Genotoxicity

Many methods ranging from comet assay to the HPRT gene mutation test have found that metal based NPs disrupt DNA and its replication process in a variety of cells. In a study examining the effects of nanosilver on DNA, AgNPs were introduced to lymphocyte cell DNA which was then examined for abnormalities. The exposure of the NPs correlated to a significant increase in micronuclei indicative of genetic fragmentation. Metal Oxides such as copper oxide, uraninite, and cobalt oxide have also been found to exert significant stress on exposed DNA. The damage done to the DNA will often result in mutated cells and colonies as found with the HPRT gene test.

Coatings and Charges

NPs, in their implementation, are covered with coatings and sometimes given positive or negative charges depending upon the intended function. Studies have found that these external factors affect the degree of toxicity of NPs. Positive charges are usually found to amplify and cause of cellular damage much more noticeably than negative charges do. In a study in which b- and c-polyethylenimine coated AgNPs were attached to strands of Lambda DNA, the cationic b-polyethylenimine coated AgNP was found to lower the melting point of the DNA 50 °C lower than its anionic counterpart.

Immunogenicity of Nanoparticles

Very little attention has been directed towards the potential immunogenicity of nanostructures. Nanostructures can activate the immune system, inducing inflammation, immune responses, allergy, or even affect to the immune cells in a deleterious or beneficial way (immunosuppression in autoimmune diseases, improving immune responses in vaccines). More studies are needed in order to know the potential deleterious or beneficial effects of nanostructures in the immune system. In comparison to conventional pharmaceutical agents, nanostructures have very large sizes, and immune cells, especially phagocytic cells, recognize and try to destroy them.

Complications with Nanotoxicity Studies

Size is therefore a key factor in determining the potential toxicity of a particle. However it is not the only important factor. Other properties of nanomaterials that influence toxicity include: chemical composition, shape, surface structure, surface charge, aggregation and solubility, and the presence or absence of functional groups of other chemicals. The large number of variables influencing toxicity means that it is difficult to generalise about health risks associated with exposure to nanomaterials – each new nanomaterial must be assessed individually and all material properties must be taken into account.

In addition, standarization of toxicology tests between laboratories are needed. Díaz, B. *et al.* from the University of Vigo (Spain) has shown (Small, 2008) that many different cell lines should be studied in order to know if a nanostructure induces toxicity, and human cells can internalize aggregated nanoparticles. Moreover, it is important to take into account that many nanostructures aggregate in biological fluids, but groups manufacturing nanostructures do not care much about this matter. Many efforts of interdisciplinary groups are strongly needed in order to progress in this field.

Effect of Aggregation or Agglomeration of Nanoparticles

Many nanoparticles agglomerate or aggregate when they are placed in environmental or biological fluids. The terms agglomeration and aggregation have distinct definitions according to the standards organizations ISO and ASTM, where agglomeration signifies more loosely bound particles and aggregation signifies very tightly bound or fused particles (typically occurring during synthesis or drying). Nanoparticles frequently agglomerate due to the high ionic strength of environmental and biological fluids, which shields the repulsion due to charges on the nanoparticles. Unfortunately, agglomeration has frequently been ignored in nanotoxicity studies, even though agglomeration would be expected to affect nanotoxicity since it changes the size, surface area, and sedimentation properties of the nanoparticles. In addition, many nanoparticles will agglomerate to some extent in the environment or in the body before they reach their target, so it is desirable to study how toxicity is affected by agglomeration.

A method was published that can be used to produce different mean sizes of stable agglomerates of several metal, metal oxide, and polymer nanoparticles in cell culture media for cell toxicity studies. Different mean sizes of agglomerates are produced by allowing the nanoparticles to agglomerate to a particular size in cell culture media without protein, and then adding protein to coat the agglomerates and "freeze" them at that size. By waiting different amounts of time before adding protein, different mean sizes of agglomerates of a single type of nanoparticle can be produced in an otherwise identical solution, allowing one to study how agglomerate size affects toxicity. In addition, it was found that vortexing while adding a high concentration of nanoparticles to the cell culture media produces much less agglomerated nanoparticles than if the dispersed solution is only mixed after adding the nanoparticles.

The agglomeration/deagglomeration (mechanical stability) potentials of airborne engineered nanoparticle clusters also have significant influences on their size distribution profiles at the end-point of their environmental transport routes. Different aerosolization and deagglomeration systems have been established to test stability of nanoparticle agglomerates. For example, laboratory setups based on critical orifices have been used to apply a wide range of external shear forces onto airborne nanoparticles. After applying shear forces, the particle mean size decreased while the particle generation rate increased. In another pioneering study, four powder aerosolization systems (dustiness testing systems) were compared for the first time for their characteristics linked to aerosol generation.

Challenges of The Nano-visualisation and Related Unknowns in Nanotoxicology

With comparison to more conventional toxicology studies, the nanotoxicology field is however suffering from a lack of easy characterisation of the potential contaminants, the "nano" scale being a scale difficult to comprehend. The biological systems are themselves still not completely known at this scale. Ultimate Atomic visualisation methods such as Electron microscopy (SEM and TEM) and Atomic force microscopy (AFM) analysis allow visualisation of the nano world. Further nanotoxicology studies will require precise characterisation of the specificities of a given nano-element : size, chemical composition, detailed shape, level of aggregation, combination with other vectors, etc. Above all, these properties would have to be determined not only on the nanocomponent before its introduction in the living environment but also in the (mostly aqueous) biological environment.

There is a need for new methodologies to quickly assess the presence and reactivity of nanoparticles in commercial, environmental, and biological samples since current detection techniques require expensive and complex analytical instrumentation. There have been recent attempts to address these issues by developing and investigating sensitive, simple and portable colorimetric detection assays that assess for the surface reactivity of NPs, which can be used to detect the presence of NPs, in environmental and biological relevant samples. Surface redox reactivity is a key emerging property related to potential toxicity of NPs with living cells, and can be used as a key surrogate for determine for the presence of NPs and a first tier analytical strategy toward assessing NP exposures.

It is difficult to determine the degree of effect of a specific nanoparticle when compared to those of comparable nanoparticles already present in our natural environment .

AEM - Analytical Electron Microscopy was used over 40 years ago to investigate amphibole asbestos bodies in Lake Superior from the Reserve Mining operations. This could non-destructively characterise sub-micron particles. Today AEM can fully characterise to atom dimensions.

Nanomaterials Pollution

Nanopollution is a generic name for waste generated by nanodevices or during the nanomaterials manufacturing process. Ecotoxicological impacts of nanoparticles and the potential for bioaccumulation in plants and microorganisms is a subject of current research, as nanoparticles are considered to present novel environmental impacts. Of the US$710 million spent in 2002 by the U.S. government on nanotechnology research, $500,000 was spent on environmental impact assessments.

Groups opposing the installation of nanotechnology laboratories in Grenoble, France, spraypainted their opposition on a former fortress above the city in 2007.

Overview

Nanowaste is mainly the group of particles that are released into the environment, or the particles that are thrown away when still on their products. The thrown away nanoparticles are usually still functioning how they are supposed to (still have their individual properties), they are just not being properly used anymore. Most of the time, they are lost due to contact with different environments. Silver nanoparticles, for example, they are used a lot in clothes to control odor, those particles are lost when washing them. The fact that they are still functioning and are so small is what makes nanowaste a concern. It can float in the air and might easily penetrate animal and plant cells causing unknown effects. Due to its small size, nanoparticles can have different properties than their own material when on a bigger size, and they are also functioning more efficiently because of its greater surface area. Most human-made nanoparticles do not appear in nature, so living organisms may not have appropriate means to deal with nanowaste.

The capacity for nanoparticles to function as a transport mechanism also raises concern about the transport of heavy metals and other environmental contaminants. Two areas of concern can be identified. First, in their free form nanoparticles can be released into the air or water during production, or production accidents, or as waste by-product of production, and ultimately accumulate in the soil, water, or plant life. Second, in fixed form, where they are part of a manufactured substance or product, they will ultimately have to be recycled or disposed of as waste.

Scrinis raises concerns about nano-pollution, and argues that it is not currently possible to "precisely predict or control the ecological impacts of the release of these nano-products into the environment." A May 2007 Report to the UK Department for Environment, Food and Rural Affairs noted concerns about the toxicological impacts of nanoparticles in relation to both hazard and exposure. The report recommended comprehensive toxicological testing and independent performance tests of fuel additives. Risks have been identified by Uskokovic in 2007. Concerns have also been raised about Silver Nano technology used by Samsung in a range of appliances such as washing machines and air purifiers.

One already known consequences to metals exposure is shown by silver, if exposed to humans in a certain concentration, it can cause illnesses such as argyria and argyrosis. Silver can also cause some environmental problems. Due to its antimicrobial properties (antibacterial), when encountered in the soil it can kill beneficial bacteria that are important to keep the soil healthy. Environmental assessment is justified as nanoparticles present novel environmental impacts. Scrinis raises concerns about nano-pollution, and argues that it is not currently possible to "precisely predict or control the ecological impacts of the release of these nano-products into the environment."

Metals, in particular, have a really strong bonds. Their properties follow up to the nanoscale as well. Metals can stay and damage the environment for a long time, since they hardly degrade or get destroyed. With the increase in use of nanotechnology, it is predicted that the nanowaste of metals will keep increasing, and until a solution is found for that problem, that waste will keep accumulating in the environment. On the other hand, some possible future applications of nanotechnology have the potential to benefit the environment. Nanofiltration, based on the use of membranes with extremely small pores smaller than 10 nm (perhaps composed of nanotubes) are suitable for a mechanical filtration for the removal of ions or the separation of different fluids. A couple of studies have found a solution to filtrate and extract those nanoparticles from water. The process is still being studied but simulations have been giving a total of about 90% to 99% removal of nanowaste particles from the water at an upgraded waste water treatment plant. Once the particles are separated from the water, they go to the landfill with the rest of the solids. Furthermore, magnetic nanoparticles offer an effective and reliable method to remove heavy metal contaminants from waste water. Using nanoscale particles increases the efficiency to absorb the contaminants and is comparatively inexpensive compared to traditional precipitation and filtration methods. One current method to recover nanoparticles is the Cloud Point Extraction. With this technique, gold nanoparticles and some other types of particles that are heat conductors are able to be extracted from aqueous solutions. The process consists of a heating section of the solution that contains the nanoparticles, and then centrifuged in order to separate the layers and then separate the nanoparticles.

Life Cycle Responsibility

To properly assess the health hazards of engineered nanoparticles the whole life cycle of these particles needs to be evaluated, including their fabrication, storage and distribution, application and potential abuse, and disposal. The impact on humans or the environment may vary at different stages of the life cycle.

The Royal Society report identified a risk of nanoparticles or nanotubes being released during disposal, destruction and recycling, and recommended that "manufacturers of products that fall under extended producer responsibility regimes such as end-of-life regulations publish procedures outlining how these materials will be managed to minimize possible human and environmental exposure" (p.xiii). Reflecting the challenges for ensuring responsible life cycle regulation, the Institute for Food and Agricultural Standards has proposed standards for nanotechnology research and development should be integrated across consumer, worker and environmental standards. They also propose that NGOs and other citizen groups play a meaningful role in the development of these standards.

Regulation of Nanotechnology

Because of the ongoing controversy on the implications of nanotechnology, there is significant debate concerning whether nanotechnology or nanotechnology-based products merit special government regulation. This mainly relates to when to assess new substances prior to their release into the market, community and environment.

Nanotechnology refers to an increasing number of commercially available products – from socks and trousers to tennis racquets and cleaning cloths. Such nanotechnologies and their accompanying industries have triggered calls for increased community participation and effective regulatory arrangements. However, these calls have presently not led to such comprehensive regulation to oversee research and the commercial application of nanotechnologies, or any comprehensive labeling for products that contain nanoparticles or are derived from nano-processes.

Regulatory bodies such as the United States Environmental Protection Agency and the Food and Drug Administration in the U.S. or the Health and Consumer Protection Directorate of the European Commission have started dealing with the potential risks posed by nanoparticles. So far, neither engineered nanoparticles nor the products and materials that contain them are subject to any special regulation regarding production, handling or labelling.

Managing Risks: Human and Environmental Health and Safety

Studies of the health impact of airborne particles generally shown that for toxic materials, smaller particles are more toxic. This is due in part to the fact that, given the same mass per volume, the dose in terms of particle numbers increases as particle size decreases.

Based upon available data, it has been argued that current risk assessment methodologies are not suited to the hazards associated with nanoparticles; in particular, existing toxicological and eco-toxicological methods are not up to the task; exposure evaluation (dose) needs to be expressed as quantity of nanoparticles and/or surface area rather than simply mass; equipment for routine detecting and measuring nanoparticles in air, water, or soil is inadequate; and very little is known about the physiological responses to nanoparticles.

Regulatory bodies in the U.S. as well as in the EU have concluded that nanoparticles form the potential for an entirely new risk and that it is necessary to carry out an extensive analysis of the risk. The challenge for regulators is whether a matrix can be developed which would identify nanoparticles and more complex nanoformulations which are likely to have special toxicological properties or whether it is more reasonable for each particle or formulation to be tested separately.

The International Council on Nanotechnology maintains a database and Virtual Journal of scientific papers on environmental, health and safety research on nanoparticles. The database currently has over 2000 entries indexed by particle type, exposure pathway and other criteria. The Project on Emerging Nanotechnologies (PEN) currently lists 807 products that manufacturers have voluntarily identified that use nanotechnology. No labeling is required by the FDA so that number could be significantly higher. "The use of nanotechnology in consumer products and industrial applications is growing rapidly, with the products listed in the PEN inventory showing just the tip of the iceberg" according to PEN Project Director David Rejeski . A list of those products that have been voluntarily disclosed by their manufacturers is located here .

The Material Safety Data Sheet that must be issued for certain materials often does not differentiate between bulk and nanoscale size of the material in question and even when it does these MSDS are advisory only.

Democratic Governance

Many argue that government has a responsibility to provide opportunities for the public to be involved in the development of new forms of science and technology. Community engagement can be achieved through various means or mechanisms. An online journal article identifies traditional approaches such as referenda, consultation documents, and advisory committees that include community members and other stakeholders. Other conventional approaches include public meetings and "closed" dialog with stakeholders. More contemporary engagement processes that have been employed to include community members in decisions about nanotechnology include citizens' juries and consensus conferences. Leach and Scoones (2006, p. 45) argue that since that "most debates about science and technology options involve uncertainty, and often ignorance, public debate about regulatory regimes is essential."

It has been argued that limited nanotechnology labeling and regulation may exacerbate potential human and environmental health and safety issues associated with nanotechnology, and that the development of comprehensive regulation of nanotechnology will be vital to ensure that the potential risks associated with the research and commercial application of nanotechnology do not overshadow its potential benefits. Regulation may also be required to meet community expectations about responsible development of nanotechnology, as well as ensuring that public interests are included in shaping the development of nanotechnology.

Community education, engagement and consultation tend to occur "downstream": once there is at least a moderate level of awareness, and often during the process of disseminating and adapting technologies. "Upstream" engagement, by contrast, occurs much earlier in the innovation cycle and involves: "dialogue and debate about future technology options and pathways, bringing the often expert-led approaches to horizon scanning, technology foresight and scenario planning to involve a wider range of perspectives and inputs." Daniel Sarewitz Director of Arizona State University's Consortium on Science, Policy and Outcomes, argues that "by the time new devices reach the stage of commercialization and regulation, it is usually too late to alter them to correct problems." However, Xenos, et al. argue that upstream engagement can be utilized in this area through anticipated discussion with peers. Upstream engagement in this sense is "meant to create the best possible conditions for sound policy making and public judgments based on carefull assessment of objective information". Discussion may act as a catalyst for upstream engagement by prompting accountability for individuals to seek and process additional information ("anticipatory elaboration"). However, though anticipated discussion did lead to participants seeking further information, Xenos et al. found that factual information was not primarily sought out; instead, individuals sought out opinion pieces and editorials.

The stance that the research, development and use of nanotechnology should be subject to control by the public sector is sometimes referred to as nanosocialism.

Newness

The question of whether nanotechnology represents something 'new' must be answered to decide how best nanotechnology should be regulated. The Royal Society recommended that the UK government assess chemicals in the form of nanoparticles or nanotubes as new substances. Subsequent to this, in 2007 a coalition of over forty groups called for nanomaterials to be classified as new substances, and regulated as such.

Despite these recommendations, chemicals comprising nanoparticles that have previously been subject to assessment and regulation may be exempt from regulation, regardless of the potential for different risks and impacts. In contrast, nanomaterials are often recognized as 'new' from the perspective of intellectual property rights (IPRs), and as such are commercially protected via patenting laws.

There is significant debate about who is responsible for the regulation of nanotechnology. While some non-nanotechnology specific regulatory agencies currently cover some products and processes (to varying degrees) – by "bolting on" nanotechnology to existing regulations – there are clear gaps in these regimes. This enables some nanotechnology applications to figuratively "slip through the cracks" without being covered by any regulations. An example of this has occurred in the US, and involves nanoparticles of titanium dioxide (TIo2) for use in sunscreen where they create a clearer cosmetic appearance. In this case, the US Food and Drug Administration (FDA) reviewed the immediate health effects of exposure to nanoparticles of titanium dioxide (TIo2) for consumers. However, they did not review its impacts for aquatic ecosystems when the sunscreen rubs off, nor did the EPA, or any other agency. Similarly the Australian equivalent of the FDA, the Therapeutic Goods Administration (TGA) approved the use of nanoparticles in sunscreens (without the requirement for package labelling) after a thorough review of the literature, on the basis that although nanoparticles of TIo2 and zinc oxide (ZNo) in sunscreens do produce free radicals and oxidative DNA damage *in vitro*, such particles were unlikely to pass the dead outer cells of the stratum corneum of human skin; a finding which some academics have argued seemed not to apply the precautionary principle in relation to prolonged use on children with cut skin, the elderly with thin skin, people with diseased skin or use over flexural creases. Doubts over the TGA's decision were raised with publication of a paper showing that the uncoated anatase form of TIo2 used in some Australian sunscreens caused a photocatalytic reaction that degraded the surface of newly installed prepainted steel roofs in places where they came in contact with the sunscreen coated hands of workmen. Such gaps in regulation are likely to continue alongside the development and commercialization of increasingly complex second and third generation nanotechnologies.

Nanomedicines are just beginning to enter drug regulatory processes, but within a few decades could comprise a dominant group within the class of innovative pharmaceuticals, the current thinking of government safety and cost-effectiveness regulators appearing to be that these products give rise to few if any nano-specific issues. Some academics (such as Thomas Alured Faunce) have challenged that proposition and suggest that nanomedicines may create unique or heightened policy challenges for government systems of cost-effectiveness as well as safety regulation. There are also significant public good aspects to the regulation of nanotechnology, particularly with regard to ensuring that industry involvement in standard-setting does not become a means of reducing competition and that nanotechnology policy and regulation encourages new models of safe drug discovery and development more systematically targeted at the global burden of disease.

Self-regulation attempts may well fail, due to the inherent conflict of interest in asking any organization to police itself. If the public becomes aware of this failure, an external, independent organization is often given the duty of policing them, sometimes with highly punitive measures taken against the organization. The Food and Drug Administration notes that it only regulates on the basis of voluntary claims made by the product manufacturer. If no claims are made by a manufacturer, then the FDA may be unaware of nanotechnology being employed.

Yet regulations worldwide still fail to distinguish between materials in their nanoscale and bulk form. This means that nanomaterials remain effectively unregulated; there is no regulatory requirement for nanomaterials to face new health and safety testing or environmental impact assessment prior to their use in commercial products, if these materials have already been approved in bulk form. The health risks of nanomaterials are of particular concern for workers who may face occupational exposure to nanomaterials at higher levels, and on a more routine basis, than the general public.

International Law

There is no international regulation of nanoproducts or the underlying nanotechnology. Nor are there any internationally agreed definitions or terminology for nanotechnology, no internationally agreed protocols for toxicity testing of nanoparticles, and no standardized protocols for evaluating the environmental impacts of nanoparticles.

Since products that are produced using nanotechnologies will likely enter international trade, it is argued that it will be necessary to harmonize nanotechnology standards across national borders. There is concern that some countries, most notably developing countries, will be excluded from international standards negotiations. The Institute for Food and Agricultural Standards notes that "developing countries should have a say in international nanotechnology standards development, even if they lack capacity to enforce the standards". (p. 14).

Concerns about monopolies and concentrated control and ownership of new nanotechnologies were raised in community workshops in Australia in 2004.

Arguments Against Regulation

Wide use of the term nanotechnology in recent years has created the impression that regulatory frameworks are suddenly having to contend with entirely new challenges that they are unequipped to deal with. Many regulatory systems around the world already assess new substances or products for safety on a case by case basis, before they are permitted on the market. These regulatory systems have been assessing the safety of nanometre scale molecular arrangements for many years and many substances comprising nanometre scale particles have been in use for decades e.g. Carbon black, Titanium dioxide, Zinc oxide, Bentonite, Aluminum silicate, Iron oxides, Silicon dioxide, Diatomaceous earth, Kaolin, Talc, Montmorillonite, Magnesium oxide, Copper sulphate.

These existing approval frameworks almost universally use the best available science to assess safety and do not approve substances or products with an unacceptable risk benefit profile. One proposal is to simply treat particle size as one of the several parameters defining a substance to be approved, rather than creating special rules for all particles of a given size regardless of type. A major argument against special regulation of nanotechnology is that the projected applications with the greatest impact are far in the future, and it is unclear how to regulate technologies whose feasibility is speculative at this point. In the meantime, it has been argued that the immediate applications of nanomaterials raise challenges not much different from those of introducing any other new material, and can be dealt with by minor tweaks to existing regulatory schemes rather than sweeping regulation of entire scientific fields.

A truly precautionary approach to regulation could severely impede development in the field of nanotechnology safety studies are required for each and every nanoscience application. While the outcome of these studies can form the basis for government and international regulations, a more reasonable approach might be development of a risk matrix that identifies likely culprits.

Response from Governments

United Kingdom

In its seminal 2004 report *Nanoscience and Nanotechnologies: Opportunities and Uncertainties*, the United Kingdom's Royal Society concluded that:

> *Many nanotechnologies pose no new risks to health and almost all the concerns relate to the potential impacts of deliberately manufactured nanoparticles and nanotubes that are free rather than fixed to or within a material... We expect the likelihood of nanoparticles or nanotubes being released from products in which they have been fixed or embedded (such as composites) to be low but have recommended that manufacturers assess this potential exposure risk for the lifecycle of the product and make their findings available to the relevant regulatory bodies... It is very unlikely that new manufactured nanoparticles could be introduced into humans in doses sufficient to cause the health effects that have been associated with [normal air pollution].*

but have recommended that nanomaterials be regulated as new chemicals, that research laboratories and factories treat nanomaterials "as if they were hazardous", that release of nanomaterials into the environment be avoided as far as possible, and that products containing nanomaterials be subject to new safety testing requirements prior to their commercial release.

The 2004 report by the UK Royal Society and Royal Academy of Engineers noted that existing UK regulations did not require additional testing when existing substances were produced in nanoparticulate form. The Royal Society recommended that such regulations were revised so that "chemicals produced in the form of nanoparticles and nanotubes be treated as new chemicals under these regulatory frameworks" (p.xi). They also recommended that existing regulation be modified on a precautionary basis because they expect that "the toxicity of chemicals in the form of free nanoparticles and nanotubes cannot be predicted from their toxicity in a larger form and... in some cases they will be more toxic than the same mass of the same chemical in larger form."

The Better Regulation Commission's earlier 2003 report had recommended that the UK Government:

1. enable, through an informed debate, the public to consider the risks for themselves, and help them to make their own decisions by providing suitable information;

2. be open about how it makes decisions, and acknowledge where there are uncertainties;

3. communicate with, and involve as far as possible, the public in the decision making process;

4. ensure it develops two-way communication channels;

5. take a strong lead over the handling of any risk issues, particularly information provision and policy implementation.

These recommendations were accepted in principle by the UK Government. Noting that there was "no obvious focus for an informed public debate of the type suggested by the Task Force", the UK government's response was to accept the recommendations.

The Royal Society's 2004 report identified two distinct governance issues:

1. the "role and behaviour of institutions" and their ability to "minimise unintended consequences" through adequate regulation;

2. the extent to which the public can trust and play a role in determining the trajectories that nanotechnologies may follow as they develop.

United States

Rather than adopt a new nano-specific regulatory framework, the United States' Food and Drug Administration (FDA) convenes an 'interest group' each quarter with representatives of FDA centers that have responsibility for assessment and regulation of different substances and products. This interest group ensures coordination and communication. A September 2009 FDA document called for identifying sources of nanomaterials, how they move in the environment, the problems they might cause for people, animals and plants, and how these problems could be avoided or mitigated.

The Bush administration in 2007 decided that no special regulations or labeling of nanoparticles were required. Critics derided this as treating consumers like a "guinea pig" without sufficient notice due to lack of labelling.

Berkeley, CA is currently the only city in the United States to regulate nanotechnology. Cambridge, MA in 2008 considered enacting a similar law, but the committee it instituted to study the issue Cambridge recommended against regulation in its final report, recommending instead other steps to facilitate information-gathering about potential effects of nanomaterials.

On December 10, 2008 the U.S. National Research Council released a report calling for more regulation of nanotechnology.

California

Assembly Bill (AB) 289 (2006) authorizes the Department of Toxic Substances Control (DTSC) within the California Environmental Protection Agency and other agencies to request information on environmental and health impacts from chemical manufacturers and importers, including testing techniques.

California

In October 2008, the Department of Toxic Substances Control (DTSC), within the California Environmental Protection Agency, announced its intent to request information regarding analytical test methods, fate and transport in the environment, and other relevant information from manufacturers of carbon nanotubes. DTSC is exercising its authority under the California Health and Safety Code, Chapter 699, sections 57018-57020. These sections were added as a result of the adoption of Assembly Bill AB 289 (2006). They are intended to make information on the fate and transport, detection and analysis, and other information on chemicals more available. The law places the responsibility to provide this information to the Department on those who manufacture or import the chemicals.

On January 22, 2009, a formal information request letter was sent to manufacturers who produce or import carbon nanotubes in California, or who may export carbon nanotubes into the State. This letter constitutes the first formal implementation of the authorities placed into statute by AB 289 and is directed to manufacturers of carbon nanotubes, both industry and academia within the State, and to manufacturers outside California who export carbon nanotubes to California. This request for information must be met by the manufacturers within one year. DTSC is waiting for the upcoming January 22, 2010 deadline for responses to the data call-in.

The California Nano Industry Network and DTSC hosted a full-day symposium on November 16, 2009 in Sacramento, CA. This symposium provided an opportunity to hear from nanotechnology industry experts and discuss future regulatory considerations in California.

DTSC is expanding the Specific Chemical Information Call-in to members of the nano-metal oxides. Interested individuals are encouraged to visit their website for the latest up-to-date information at http://www.dtsc.ca.gov/TechnologyDevelopment/Nanotechnology/index.cfm.

On December 21, 2010, the Department of Toxic Substances Control (DTSC) initiated the second Chemical Information Call-in for six nanomaterials: nano cerium oxide, nano silver, nano titanium dioxide, nano zero valent iron, nano zinc oxide, and quantum dots. DTSC sent a formal information request letter to forty manufacturers who produce or import the six nanomaterials in California, or who may export them into the State. The Chemical Information Call-in is meant to identify information gaps of these six nanomaterials and to develop further knowledge of their analytical test methods, fate and transport in the environment, and other relevant information under California Health and Safety Code, Chapter 699, sections 57018-57020. DTSC completed the carbon nanotube information call-in in June 2010.

DTSC partners with University of California, Los Angeles (UCLA), Santa Barbara (UCSB), and Riverside (UCR), University of Southern California (USC), Stanford University, Center for Environmental Implications of Nanotechnology (CEIN), and The National Institute for Occupational Safety and Health (NIOSH) on safe nanomaterial handling practices.

DTSC is interested in expanding the Chemical Information Call-in to members of the bominated flame retardants, members of the methyl siloxanes, ocean plastics, nano-clay, and other emerging chemicals.

European Union

The European Union has formed a group to study the implications of nanotechnology called the Scientific Committee on Emerging and Newly Identified Health Risks which has published a list of risks associated with nanoparticles.

Consequently, manufacturers and importers of carbon products, including carbon nano-tubes will have to submit full health and safety data within a year or so in order to comply with REACH.

Response from Advocacy Groups

In January 2008, a coalition of over 40 civil society groups endorsed a statement of principles calling for precautionary action related to nanotechnology. The coalition called for strong, comprehensive oversight of the new technology and its products in the International Center for Technology Assessment's report *Principles for the Oversight of Nanotechnologies and Nano materials*, which states:

> *Hundreds of consumer products incorporating nano-materials are now on the*

market, including cosmetics, sunscreens, sporting goods, clothing, electronics, baby and infant products, and food and food packaging. But evidence indicates that current nano-materials may pose significant health, safety, and environmental hazards. In addition, the profound social, economic, and ethical challenges posed by nano-scale technologies have yet to be addressed ... 'Since there is currently no government oversight and no labeling requirements for nano-products anywhere in the world, no one knows when they are exposed to potential nano-tech risks and no one is monitoring for potential health or environmental harm. That's why we believe oversight action based on our principles is urgent' ... This industrial boom is creating a growing nano-workforce which is predicted to reach two million globally by 2015. 'Even though potential health hazards stemming from exposure have been clearly identified, there are no mandatory workplace measures that require exposures to be assessed, workers to be trained, or control measures to be implemented,' explained Bill Kojola of the AFL-CIO. 'This technology should not be rushed to market until these failings are corrected and workers assured of their safety'" also.

The group has urged action based on eight principles. They are 1) A Precautionary Foundation 2) Mandatory Nano-specific Regulations 3) Health and Safety of the Public and Workers 4) Environmental Protection 5) Transparency 6) Public Participation 7) Inclusion of Broader Impacts and 8) Manufacturer Liability.

Some NGOs, including Friends of the Earth, are calling for the formation of a separate nanotechnology specific regulatory framework for the regulation of nanotechnology. In Australia, Friends of the Earth propose the establishment of a Nanotechnology Regulatory Coordination Agency, overseen by a Foresight and Technology Assessment Board. The advantage of this arrangement is that it could ensure a centralized body of experts that are able to provide oversight across the range of nano-products and sectors. It is also argued that a centralized regulatory approach would simplify the regulatory environment, thereby supporting industry innovation. A National Nanotechnology Regulator could coordinate existing regulations related to nanotechnology (including intellectual property, civil liberties, product safety, occupation health and safety, environmental and international law). Regulatory mechanisms could vary from "hard law at one extreme through licensing and codes of practice to 'soft' self-regulation and negotiation in order to influence behavior." The formation of national nanotechnology regulatory bodies may also assist in establishing global regulatory frameworks.

In early 2008, The UK's largest organic certifier, the Soil Association, announced that its organic standard would exclude nanotechnology, recognizing the associated human and environmental health and safety risks. Certified organic standards in Australia exclude engineered nanoparticles. It appears likely that other organic certifiers will also follow suit. The Soil Association was also the first to declare organic standards free from genetic engineering.

Technical Aspects

Size

Regulation of nanotechnology will require a definition of the size, in which particles and processes are recognized as operating at the nano-scale. The size-defining characteristic of nanotechnology is the subject of significant debate, and varies to include particles and materials in the scale of at least 100 to 300 nanometers (nm). Friends of the Earth Australia recommend defining nano-particles up to 300 nanometers (nm) in size. They argue that "particles up to a few hundred nanometers in size share many of the novel biological behaviors of nano-particles, including novel toxicity risks", and that "nano-materials up to approximately 300 nm in size can be taken up by individual cells". The UK Soil Association define nanotechnology to include manufactured nano-particles where the mean particle size is 200 nm or smaller. The U.S. National Nanotechnology Initiative defines nanotechnology as "the understanding and control of matter at dimensions of roughly 1 to 100 nm.

Mass Thresholds

Regulatory frameworks for chemicals tend to be triggered by mass thresholds. This is certainly the case for the management of toxic chemicals in Australia through the National pollutant inventory. However, in the case of nanotechnology, nano-particle applications are unlikely to exceed these thresholds (tonnes/kilograms) due to the size and weight of nano-particles. As such, the Woodrow Wilson International Center for Scholars questions the usefulness of regulating nanotechnologies on the basis of their size/weight alone. They argue, for example, that the toxicity of nano-participles is more related to surface area than weight, and that emerging regulations should also take account of such factors.

References

- Chris Phoenix; Mike Treder (2008). "Chapter 21: Nanotechnology as global catastrophic risk". In Bostrom, Nick; Cirkovic, Milan M. Global catastrophic risks. Oxford: Oxford University Press. ISBN 978-0-19-857050-9

- "About the National Nanotechnology Initiative". United States National Nanotechnology Initiative. 2016. Retrieved 16, June 2020

- Smith, Erin Geiger (14 February 2013). "U.S.-based inventors lead world in nanotechnology patents: study". Technology. Reuters. Retrieved 04, June 2020

- Chan WCW (2007). "Toxicity Studies of Fullerenes and Derivatives". Bio-applications of nanoparticles. New York, NY: Springer Science + Business Media. ISBN 0-387-76712-6

Permissions

All chapters in this book are published with permission under the Creative Commons Attribution Share Alike License or equivalent. Every chapter published in this book has been scrutinized by our experts. Their significance has been extensively debated. The topics covered herein carry significant information for a comprehensive understanding. They may even be implemented as practical applications or may be referred to as a beginning point for further studies.

We would like to thank the editorial team for lending their expertise to make the book truly unique. They have played a crucial role in the development of this book. Without their invaluable contributions this book wouldn't have been possible. They have made vital efforts to compile up to date information on the varied aspects of this subject to make this book a valuable addition to the collection of many professionals and students.

This book was conceptualized with the vision of imparting up-to-date and integrated information in this field. To ensure the same, a matchless editorial board was set up. Every individual on the board went through rigorous rounds of assessment to prove their worth. After which they invested a large part of their time researching and compiling the most relevant data for our readers.

The editorial board has been involved in producing this book since its inception. They have spent rigorous hours researching and exploring the diverse topics which have resulted in the successful publishing of this book. They have passed on their knowledge of decades through this book. To expedite this challenging task, the publisher supported the team at every step. A small team of assistant editors was also appointed to further simplify the editing procedure and attain best results for the readers.

Apart from the editorial board, the designing team has also invested a significant amount of their time in understanding the subject and creating the most relevant covers. They scrutinized every image to scout for the most suitable representation of the subject and create an appropriate cover for the book.

The publishing team has been an ardent support to the editorial, designing and production team. Their endless efforts to recruit the best for this project, has resulted in the accomplishment of this book. They are a veteran in the field of academics and their pool of knowledge is as vast as their experience in printing. Their expertise and guidance has proved useful at every step. Their uncompromising quality standards have made this book an exceptional effort. Their encouragement from time to time has been an inspiration for everyone.

The publisher and the editorial board hope that this book will prove to be a valuable piece of knowledge for students, practitioners and scholars across the globe.

Index

A
Adenine, 66, 70, 159
Amino Acids, 168

B
Binary Code, 153
Biological Membranes, 179, 217
Biosynthesis, 70
Block Copolymers, 180
Blood Pressure, 215
Bottom-up Approach, 60, 112, 117, 124, 179
Brownian Motion, 92, 98, 101
Bulk Materials, 1, 4, 10, 12, 26, 40, 57, 66, 87, 111, 134, 136

C
Carbon Black, 61, 215, 229
Carbon Monoxide, 44-45
Carbon Nanotubes, 2-3, 5, 9, 18, 114, 137, 141, 153, 161, 163, 179, 183, 187, 194, 198, 207, 211, 215, 219, 232
Carrier Gas, 67, 148
Catalytic Activity, 43
Cathode, 85, 106
Cell Culture, 221
Cell Cycle, 81
Cell Nucleus, 217
Chain Length, 176
Chemical Reaction, 16, 98, 164
Chemical Resistance, 178
Chemical Vapor Deposition, 65, 102, 123-124, 153, 191, 193
Chirality, 196
Chlorine, 74, 125
Commercialization, 104, 154, 227-228
Communication Channels, 231
Composite Materials, 65, 137, 191, 199-200
Composition, 22, 66, 71, 75, 104, 127, 149, 173, 188, 192, 206, 221
Conduction Band, 36-37, 71, 84, 87-90, 105-106
Conjunction, 152
Covalent Binding, 163
Covalent Bonding, 13
Covalent Bonds, 76, 172
Cryo-electron Microscopy, 169
Cytosine, 159

D
Debye Length, 118
Diffusion Process, 176
Diodes, 104
Disruptive Technology, 108
Dna Double Helix, 158, 179
Dna Nanotechnology, 157-160, 162-166, 169-170
Dna Replication, 169

L
Labor Markets, 209
Laser Ablation Method, 192
Ligand, 34, 90
Light Spectrum, 148
Logic Gate, 163
Logic Gates, 116, 165

M
Magnesium, 28, 229
Magnetic Field, 18, 79, 194, 199
Magnetic Nanoparticles, 78, 178, 224

N
Nanoimprint Lithography, 178
Nanolithography, 176-177, 179
Nanomedicine, 73, 76, 157, 166, 204-205
Nanomesh, 120-123
Nanometer Scale, 73, 111
Nanoparticle, 12-13, 17, 20, 32-34, 37-39, 43, 56, 58, 71, 77, 80, 82, 98, 132, 166, 173, 178, 193, 204, 215, 217, 221

Nanoparticle Toxicity, 217
Nanopatterning, 64
Nanopore Sequencing, 149-150, 153, 155
Nanorods, 52, 69, 71, 73, 89-90, 126
Nanostructures, 2, 46, 48-49, 57, 64-66, 71, 76, 79, 89, 115, 118, 125, 127, 130, 134, 138, 141, 158, 162, 166, 168, 177, 182, 204, 220
Nanowires, 71, 90, 108, 111-119, 135-137, 141, 143-146, 148, 182-183
Nervous System, 217
Nicotinamide Adenine Dinucleotide, 66, 70
Nitrate, 28, 30, 70
Nucleic Acid, 157-160, 162-170, 183
Nucleic Acids, 157-160, 162, 168, 170, 182

O
Oligonucleotide Synthesis, 168
Organic Compounds, 93
Organic Matter, 201

P
Pathogenic Bacteria, 71
Photon, 42, 57-58, 84-85, 87, 103, 116
Photonic Crystals, 108, 124
Photonics, 108
Plasmonic Nanoparticles, 80
Polymer, 6, 21, 94, 106, 123, 130, 132, 166, 178, 199, 202, 221
Polymerization, 92, 101, 124
Polymers, 65, 68-69, 92, 105, 113, 117
Polynomials, 49
Positron Emission, 200
Positron Emission Tomography, 200
Precipitation, 43, 94, 119, 224
Proliferation, 219
Protons, 11, 74, 182
Pseudomonas, 71

T
Thymine, 159
Tile-based Structures, 158, 167, 170
Tissue Engineering, 199
Titanium Dioxide, 29, 213, 215, 228-229, 233
Topography, 124
Trajectories, 231

W
Wastewater Treatment, 201
Water Purification, 212

X
X-ray Crystallography, 157, 169
X-ray Diffraction, 190